Organic Farming
Concepts, Applications and Advances

The Authors

Dr. Subhash Chand is working as an Assistant Professor of Soil Science, Division of NRM, at FOA Wadura Campus, Sopore, of Sher-e-Kashmir University of Agricultural Sciences and Technology of Kashmir Shalimar, Srinagar, J&K. He is member of Editorial Boards of Several Scientific Journals and has published papers in scientific and technical journals. He is member of Scientific Societies *viz.* Indian Science Congress Association- Kolkata, Soil Conservation Society of India-New Delhi, Indian Society of Remote Sensing-Dehradun, Indian Society Soil Survey and Land Use Planning-Nagpur, Association of Soil & Water Conservationist-Dehradun and Indian Society of Soil Science, New Delhi. He has 12 years of experience in Research and Teaching of UG, PG and Ph.D. students. He co-chaired Technical session and parallel session at International workshop.

Sartaj Ahmad Wani received his B. Sc. (Ag.) Hons. Degree in 2010 and M. Sc. (Ag.) in Soil Science in 2012 from Sher-e-Kashmir University of Agricultural Sciences and Technology of Kashmir Shalimar, Srinagar, respectively. He is NET (ICAR) qualified and currently persuing Doctoral Research with SRF in the Soil Science, Department in the University. He has published number of research papers in journals of repute and participated in various national conferences symposiums.

Organic Farming
Concepts, Applications and Advances

Dr. Subhash Chand
Assistant Professor
Division of Natural Resource Management
SKUAST-Kashmir, J&K,
Campus Wadura, Sopore-193201
J&K, India

Sartaj A. Wani
Division of Natural Resource Management
SKUAST-Kashmir, J&K,
Campus Wadura, Sopore-193201
J&K, India

2016
Daya Publishing House®
A Division of
Astral International Pvt. Ltd.
New Delhi - 110 002

Cataloging in Publication Data—DK
  Courtesy: D.K. Agencies (P) Ltd. <docinfo@dkagencies.com>

Subhash Chand, author.
  Organic farming : concepts, applications and advances
/ Dr. Subhash Chand, Sartaj A. Wani.
    pages cm
  Includes bibliographical references.
  ISBN 9789351247067 (Hardbound)
  ISBN 9789351308898 (International Edition)

  1. Organic farming. I. Wani, Sartaj A. (Sartaj
Ahmad), author. II. Title.

  S605.5.S83 2016   DDC 631.584   23

Published by   : **Daya Publishing House®**
                        A Division of
                        Astral International Pvt. Ltd.
                        – ISO 9001:2008 Certified Company –
                        4760-61/23, Ansari Road, Darya Ganj
                        New Delhi-110 002
                        Ph. 011-43549197, 23278134
                        E-mail: info@astralint.com
                        Website: www.astralint.com

Laser Typesetting   : **SSMG Computer Graphics, Delhi - 110 035**

Printed at   : **Thomson Press India Limited**

Preface

Challenges faced in modern agriculture due to high demand of food by ever growing population have been tremendously increased. Decreasing factor productivity and soil fertility is matter of great concern. Tackling the challenges of decreased soil fertility and organic carbon which is storehouse of energy and nutrients may be encountered by judicious use of various sources of organic nutrients viz; composting, biofertilisers, green manuring, crop residues, farmyard manure, cover crops, bio pesticides, grasses, etc. are of great importance. These sources are helpful for improving soil physical, biological and chemical properties.

This book is an account of introduction, principles, different practices, recent advances related to organic farming. A comprehensive reviews on benefits and limitations of organic farming as carried out by various research workers have been discussed in detail. The biomass energy is having great potential in the scenario of scarcity of energy. Soil Organic (somo) matter is an important factor in restoring soil fertility. An increase in the soil fertility has a direct relationsehip with socio-economic conditions of farmers, vegetable growers and orchardists. A healthy soil is always responsible for production of quality grains and fruits hence there is an urgent need to maintain for management of soil health. Organic produce are healthier and tasty than inorganic foods. However, the decline in yield has been noticed and recorded various by research workers. The book *"Organic Farming Concepts, Applications and Advances"* has been written in concise form.

Authors thank to various faculty members and students for healthy discussion and feedback on organic farming. It will be useful for students, progressive farmers, research scholars, faculty members and policy planners, socio-economists, environmental experts, research institutions, colleges, universities, scientist, professor, doctors, watering mans, food scientists, cook, seruents, growth, vegetable farmers, orchardists, etc.

Dr. Subhash Chand

Sartaj A. Wani

Preface

Contents

Chapter 1

Introduction

The concept of organic farming was started 1,000 years back when ancient farmers started cultivation near the river belt depending on natural resources only. There is brief mention of several organic inputs in Indian ancient literature like *Rig-Veda,Ramayana, Mahabharata* and Kautilya *Arthasashtra* etc. In fact, organic agriculture has its roots in traditional farming practices that evolved in countless villages and farming communities over the millennium. Organic farming is an old age practice dating back to Neolithic age practiced by ancient civilization like Mesopotamia, Hwang Ho Basin etc. The scripts of Ramayana describes that all dead things –rotting corpse or stinking garbages returned to earth are transformed into wholesome thing that nourishes life on earth. Such is the alchemy of mother earth that is interpreted by C. Rajagopalachari. The *Mahabharat* mentions Kamamdhenu, the celestial cow and its role in human life and soil fertility. Kauatiya *Artharashtra* (3000) mentioned manures oil cakes, excreta of animals. *Brihad Sanahita* by Varamihir described how to choose manures of different crops and methods of manuring. *Rig Veda* (2500- 1500) mentions of organic manure in *Rig Veda Sukta* 1.161, 10. Similarly green manure in Atharva Veda II (1000BC) 8.3 in Sukta (IV, V, 94, 107- 112) it is stated for healthy growth the plant should be nourished by dung of cow, sheep, goat, water and meat. A number of studies have revealed the importance to organic farming systems in the present era for sustainable development of human existence.

Before 19th century most food in the world was organically produced using organic manures and human and animal power (horses in US and oxen in Asia) (White, 1970). The agricultural revolution in England began in the early 19th century when 'Jethro Tull' invented a horse drawn hoe and a seed drill with tines at right distance to sow the row crops. German Chemist Fritz Haber developed the process of ammonia synthesis (Prasad, 2003) which led to the manufacture of nitrogen fertilizer in US (Collings, 1955). Fertilizer N was needed in large amounts to benefit from the discovery of high yielding hybrid corn (maize). Insecticidal property of DDT was discovered in 1939 by P. Muller in Switzerland and was followed by the discovery of BHC in France and BHC in UK (Brown, 1951). Nitrophenols were the first group of selective herbicides developed in 1933 and were followed by the development of 2, 4-D and MCPA in 1940's (Rao, 1983).Thus by the middle of 20th century most of the components of the modern agriculture i.e. tractors and associated farm machines, fertilizer and agrochemicals were in the use on the agricultural farms in the developed world (Prasad, 2005).

In post-world war era, the green revolution launched in Mexico with private funding from the US, encouraged the development of hybrid plants, chemical controls, large-scale irrigation, and heavy mechanization around the world. Although science tended to concentrate on the new chemical approaches but sustainable agriculture was topic of interest. In US, J.I. Rodale (1950) began to popularise the term and methods of organic growing, particularly through promotion of organic gardening. In 1962, Rachel Carson, a prominent scientist and naturalist, published *Silent Spring*, describing the effect of DDT and other pesticides on the environment by launching the worldwide environmental movement. Lady Eve Balfour launched the Haughley experiment in 1939 in England and her publication The Living Soil, led to the formation of the Soil Association, a key international organic advocacy group. Japanese microbiologist, Masanobu Fukuoka, devoted his 60 years towards developing a radical no-till organic method for growing grain and other crops, now known as nature farming or Fukuoka farming. The international campaign of Green Revolution launched in Mexico in 1944 with private funding from the US encouraged the development and use of hybrid plants, chemical controls, large-scale irrigation, and heavy mechanization in agriculture around the world. By 1970s global movements concerned with pollution and the environment increased their focus on organic farming. Fukuoka released his first book (1975), *One Straw Revolution* with a wide ranging impact on the agricultural world. From 1980 to 2000, various consumer groups began seriously pressurising for government regulation of organic production.

The roots of organic farming can be traced to the Europe back to the first quarter of the early 20th century (Stockdale *et al.*, 2001). In 1924, the Austrian philosopher Dr. Rudolf Steiner conceptualised and advocated organic agriculture and in 1927 a trademark "Demeter" was introduced for organic food produced. Organic farming or natural farming has no doubt emerged from Asian countries like India and China where agriculture was the main stay and farmers have nurtured and groomed this art over several centuries (Subhash, 2005). In the beginning of twentieth century, there was lack of knowledge about organic agriculture. From 1905 to 1924, British botanist, Sir Albert Howard, often referred as the Father of modern organic agriculture worked as an agriculture advisor in Pusa, Samastipur, India, where he documented traditional Indian farming practices and came to regard them as superior to his conventional agriculture. His research and further developments of these methods was recorded in his book on *Agricultural Testament*, which influenced many scientists and farmers of the day. In 1939, Lady Eve Balfour lunched the *Haughley Experiment* on farmland in England. It was the first scientific comparison of organic and conventional farming. She published the *Living Soil*, based on the initial findings of the *Haughley Experiment*. In Germany (1940), Rudolf Steiner's development of biodynamic agriculture was probably the first comprehensive organic farming system. The farm organic farming is usually credited to Lord

Northbourn in his book, *Look to the Land*, in which he described a holistic, ecological balanced approach for organic farming. In Japan (1940), Masanobu Fukoka, a soil scientist working on microbiology and plant pathology, began to doubt the modern agricultural movement. He devoted the next 30 years in developing radical-no-till organic methods for growing grain, now known as *Fukuoka Farming*.

Modern agriculture has been of great help in alleviating hunger from the world, because the world population more than doubled itself during the last half of the 20th century (Lal, 2000). However, even now globally almost 9 billion people still are at the bank of hunger. The famines and scarcities have been known in India from the earliest times (Randhawa, 1983). This was all in the era of organic agriculture. As a contrast, there was no scarcity of food after the severe droughts of 1972 and 1987 (FAI, 2004) due to intensive agriculture. India's own achievements in agricultural production after the Green Revolution that set in 1967-68 has been exemplary and mainly due to increased use of the components of modern agriculture viz; fertilizer, pesticides, high yielding varieties (HYVs) and farm machinery. Food grain production in India itself more than doubled during the post Green Revolution period with virtually no increase in net cultivated area; it increased from 95 million tonnes in 1967-68 to 209 million tonnes in 1999-00 from the same net area (140±1 million ha) (FAI, 2004).With increasing population, the cultivable land resource is shrinking day by day. To meet the food, fibre, fuel, fodder and other needs of the growing population, the productivity of agricultural land and soil health needs to be improved. Green Revolution in the post-independence era has shown path to developing countries for self-sufficiency in food but sustaining agricultural production against the finite natural resource base demands has shifted from the "resource degrading" chemical agriculture to a "resource protective" biological or organic agriculture. Green revolution technologies such as greater use of synthetic agrochemicals like fertilizers and pesticides, adoption of nutrient-responsive, high-yielding varieties of crops, greater exploitation of irrigation potentials etc. has boosted the production output in most cases. However, continuous use of these high energy inputs indiscriminately now leads to decline in production and productivity of various crops as well as deterioration of soil health and environments. The most unfortunate impact of Green Revolution Technologies (GRT) on IndianAgriculture is as follows:

- Imbalance use of fertilizers
- Dependency on synthetic chemical fertilizers due to heavy feeder varieties
- Increase in secondary & micronutrient deficiencies in soils
- Increase in pesticide use due to heavy attack of insect and pest
- Non-scientific water management and distribution
- Reduction in factor productivity
- Reduction in quality of the produce
- Extinction of gene pool (Gene depression)
- Environmental pollution
- Imbalance in social and economic status (marginal and small farmers)

Save and Sanghavi (1991) are of the view that after their intensive experiments with organic farming and narrating the results to the informed, it is time that the governments and farmers are brought around. Four crops of banana grown by the natural way on the same farm by them are compared with those produced by the conventional way. While the natural farm yielded 18 kg of banana in the first round, the conventional one gave 25 kg. 30 kg was the yield at the second round on both the farms. However, on the third round, the natural farm gave 25 kg, the conventional one yielded only 20 kg. The results on the fourth round were stunning - the plants on the conventional farm died out; but the natural ones gave 15 kg on an average. Thus, the aggregate output was 88 kg on the natural farm and 75 kg on the conventional one. While, the natural banana commanded a price of Rs. 2.50 per kg, the conventional one could fetch only Rs. 1.75 per kg. This has been the major reason for the substantial net profit (Rs. 154) earned from the cultivation of natural banana (conventional banana could get only a net profit of Rs. 26.25). The expenses incurred were Rs. 66 and Rs. 105 for the natural and conventional bananas respectively. However, a stringent cost and return analysis representing a larger sample size will be necessary to draw meaningful conclusions. It should be born in mind that the output obtained from the natural banana farm was also because of the accessibility to the inputs and expertise, which the authors happened to possess. Farmers placed in less advantageous positions may not derive such results. The price advantage to the natural organic farming products will also taper off when the increases. The environmental costs and returns have to be internalized and it is quite possible that the organic farming will prove to be a far better alternative to the conventional one. However, these aspects will have to be built into a scientific and tight economic reasoning, among others.

Kaushik (1997) analyzed the issues and policy implications in the adoption of sustainable agriculture. The concept of trades off has a forceful role to play in organic farming both at the individual and national decision making levels. Public *vis-a-vis* private benefits, current *vis-a-vis* future incomes, current consumption and future growths, etc. are very pertinent issues to be determined. The author also lists a host of other issues. While this study makes a contribution at the conceptual level, it has not attempted to answer the practical questions in the minds of the farmers and other sections of the people.

Geier (1999) is of the opinion that there is no other farming method so clearly regulated by standards and rules as organic agriculture. The organic movement has decades of experience through practicing ecologically sound agriculture and also in establishing inspection and certification schemes to give the consumers the guarantee and confidence in actuality. Organic farming reduces external inputs and it is based on a holistic approach to farming. He describes the worldwide success stories of organic farming based on the performance of important countries in the west. The magnitude of organic farming products is also mentioned. To the question of whether the organic farming can feed the world, he says that neither chemical nor organic farming systems can do it; but the farmers can.

Veeresh (1999) opines that both high technology and sustainable environment cannot go together. Organic farming is conceived as one of the alternatives to conventional agriculture in order to sustain production without seriously harming the environment and ecology. However, he says that in different countries organic farming is perceived differently. While in the advanced countries, its focus is on prevention of chemical contamination, we, in countries like India are concerned of the low soil productivity. Even the capacity to absorb fertilizers depends on the organic content of the soil. The principles of organic farming are more scientific than those of the conventional. India's productivity of many crops is the lowest in the world in spite of the increase in the conventional input use. The decline in soil nutrients, particularly in areas where the chemical inputs are increasingly being used in the absence of adequate organic matter is cited as a reason for low productivity. Doubts about the availability of massive sources of organic inputs also exist. He advocates an advance to organic farming at a reasonable pace and recommends conversion of only 70 per cent of the total cultivable area where un-irrigated farming is in vogue. This 70% supplies only 40% of our food production. While this analysis has several merits, it is more addressed to the policy makers and less to the farmers.

Sankaram (2001) is of the view that almost all benefits of high yielding varieties based farming accrue mostly in the short term and in the long term they cause adverse effects. There is an urgent need for a corrective action. The author rules out organic farming based on the absolute exclusion of fertilizers and chemicals, not only for the present, but also in the foreseeable future. There ought to be an appropriate blend of conventional farming system and its alternatives. The average yields under organic and conventional practices are almost the same and the declining yield rate over time is slightly lower in organic farming. The author also quotes a US aggregate economic model, which shows substantial decreased yields on the widespread adoption of organic farming. Decreased aggregate outputs, increased farm income and increased consumer prices are other results the model gives.

Singh (2001) recording the experiments on rice-chick pea cropping sequence using organic manure, found the yields substantially higher compared to the control group. Similar results were obtained for rice, ginger, sunflower, soybean and sesame. Farmers from Uttar Pradesh have allotted a portion of their land exclusively for organic farming found that the yields of sugarcane, rice, wheat and vegetables were lower than those under chemical farming. An Englishman, settled in Tamil Nadu, who runs an organic farm in 70 acres planted with coffee, citrus, other fruits, rice, pepper and vegetables says that he does not earn a profit and does not have confidence in organic farming. Korah Mathen recounts several problems in evolving representative and rigorous yardsticks for comparison between modern and alternative farming. Yields cannot be compared, because of monoculture nature of chemical farming with those of multi crops raised under organic/natural farming. Economic analysis is also problematic because one has to quantify the intangibles. He advocated the resource use efficiency analysis.

Save (1999) found that after three years of switching over to natural cultivation, the soil was still recovering from the after effects of chemical farming. When the soil regained its health, production increased and the use of inputs decreased. The farm, which was yielding 200 to 250 coconuts per tree, gave 350 to 400 per annum.

Rahudkar and Phate (1992) narrate the experiences of organic farming in Maharashtra. Individual farmers growing sugarcane and grapes, after using vermicompost, saw the soil fertility increased, irrigation decreased by 45 per cent and sugarcane quality improved. The authors say that net profits from both the sugarcane and grape crops are high in organic farms.

Organic farming in India is an ancient traditional wisdom since the Vedic era. Our *rishis* and *munies* doing organic farming since 17th century.In India organic movement started in Madhya Pradesh (2001) and then spread all over India. Throughout the history, the focus of agricultural research and majority of published scientific findings has been shifted to biotechnologies like genetic engineering. The rise of organic farming was driven by small, independent producers and by consumers. In recent years, explosive organic market growth has encouraged the participation of agribusiness interest. In India, organic farming has started simultaneously from two streams. While the commercial growers of spices, basmati rice and cotton adopted organic for premium prices in export market, resource-poor farmers in rainfed marginal lands adopted it as an alternative livelihood approach, which not only promises clean environment and healthy food but also ensures soil fertility, long-term sustainability and freedom form debt and market forces. The renewed interest in organic farming in India is mainly due to three main reasons, reduction in agricultural yield in certain areas as a result of excessive and indiscriminate use of chemical inputs, decreased soil fertility and a concern regarding environment. The 10th Five-Year Plan encouraged the promotion of organic farming using organic wastes, and integrated pest management (IPM) and integrated nutrient management (INM) practices (GoI, 2008). Even the 9th five-year plan had emphasised the promotion of organic produce in plantation crops, spices and condiments using organic and bio-inputs for the protection of environment and promotion of sustainable agriculture. Presently, many states and private agencies are involved in the promotion of organic farming in India; these also include several ministries and government departments at both central and state levels.

Milestone in Organic Farming

Sir Albert Howard(1900-1947)	Father of modern Organic Agriculture, developed organic composting (mycorrhizal fungi) at Pusa, Samastipur, India and Published document 'and Agriculture Testament'.
Rudolph Steiner (1992)	A German spiritual Philosophical built biodynamic farm in Germany.
J. I Rodel (1950)	Popularised the term sustainable agriculture and method of organic growing in USA
IFO	AM Establishment of International Federation of organic Agriculture Movement in 1972 in France.

One Straw Revolution   Release of the book by MASANOBU Fukoka(1975),
                                     an eminent microbiologist of Japan EU regulation
                                     Codex guidance on organic standard 1999.

1.1 Scenario of Organic Farming

As per the latest survey conducted by International Federation of Organic Agriculture Movement (IFOAM) and SOEL Association (2009), almost 31 million hectares (m ha) are currently managed organically by more than 600,000 farmers worldwide and more than 62 million areas are certified. This constitutes 0.7 per cent of the agricultural land of these countries according to the 2009 survey. Countries with most organic land are Australia / Oceania with 11.9 m. ha, followed by Europe with 7 m ha. Latin America (5.8 m. ha), Asia (2.9 m. ha), North America (2.2 m. ha) and Africa (0.9 m. ha). Currently countries with more organic lands are Australia (11.8 m. ha), Argentina (3.1 m. ha), China (2.3 m. ha) and US (1.6 m. ha).

The number of farms and the proportion of organically managed land compared to conventionally managed one is highest in Europe. There has been major growth of organic area in North America and Europe. Both have added over half a million ha each during 2005-06. In North America, it represents an increase of almost 30%, an exceptional growth. In most other countries, organic farming is on the rise. There are also some decreases of organic land (extensive pastoral land) in China, Chile and Australia. As per 2007 survey of IFOAM, land use information was available for 27 million hectares. More than half of the organic agricultural land is used for permanent pasture/grassland; one quarter is used for arable cropping10% for permanent crops, followed by other crops (5%) and other land use (1%). On a global level, permanent pastures/grassland (19.8 m ha) account for almost two third of the world's organic land. More than half of this grassland is in Australia. Furthermore, large areas of permanent pastures are in Latin America and Europe. The largest collection areas are in European Africa (almost 27 m ha each). In terms of quantities, the important wild collected products are: bamboo shoot (36%), fruits and berries (21%) and nuts (19%). The main crop categories for arable land are cereals followed by fodder crops, other arable crops, set-aside/green manuring, protein crops, vegetables, oilseeds, industrial crops, medicinal and aromatic plans, root crops, seed production etc. Besides the above, there is about 62 m ha of organic wild collection area with 979 organic wild collection projects, world over.

Currently, India ranks 33rd in terms of total land under organic cultivation and 88th in agricultural land under organic crops to total farming area. According to the Agricultural and Processed Food Product Export Development Authority (APEDA), the cultivated land under certification is around 2.8 M ha (2007-08), which includes one million hectares under cultivation and the rest is under forest area (wild collection). An estimated 69 M ha, however, is traditionally cultivated without using chemical fertilizers and could be eligible for certification under the current practices, or with small modifications. Certifying these farms remains a challenge, however, as many of these farms are small holdings (nearly 60% of all farms in India are less than one ha). Small holders and resource-poor farmers may not be able to afford the cost of certification, they are illiterate and unable to maintain necessary records, or may be using indigenous cultivation systems not recognized in organic certification systems. These farms mainly produce for home consumption, and supply to the local markets in case of irregular surpluses. Such barriers pose difficulties for farms to reap potential benefits of organic certification (Reddy, B. S. 2010). As per the review by Roy Chowdhury *et al.* (2013) the percentage of area under organic farming in the total cultivated area of different countries of the world in year 2004 has been provided in table 1.1 below.

Table 1.1: Percentage of area under organic farming in the total cultivated area of different countries of the world (Roy chowdhury *et al.*, 2013).

Country	Percentage of area under organic farming
USA	0.23
UK	4.22
Germany	4.10
Argentina	1.70
Austria	8.40
Australia	2.20
Japan	0.10
Switzerland	7.94
South Africa	0.05
Italy	3.70
India	0.03
Pakistan	0.08
Sri Lanka	0.05

Chapter 2

Concept of Organic Farming

It is exclusively known for non-use of synthetic/ artificial chemicals (fertilizer, pesticides, herbicides, growth hormones etc). It is also known farming carbon because of heavy use of carbonaceous materials as a source of energy and plant nutrients. The approach and outlook towards agriculture and marketing of food has seen a quantum change worldwide over the last few decades. Whereas earlier the seasons and the climate of an area determined what would be grown and when, today it is the "market" that determines what it wants and what should be grown. The focus is now more on quantity and "outer" quality (appearance) rather than intrinsic or nutritional quality, also called "vitality".

This immense commercialisation of agriculture has also had a very negative effect on the environment. The use of pesticides has led to enormous levels of chemical build up in our environment, in soil, water, air, in animals and even in our own bodies. Fertilisers have a short-term effect on productivity but a longer-term negative effect on the environment where they remain for years after leaching and running off, contaminating ground water and water bodies. The use of hybrid seeds and the practice of monoculture have led to a severe threat to local and indigenous varieties, whose germplasm can be lost forever.

In the name of growing more to feed the earth, we have taken the wrong road of unsustainability. The effects already show-farmers committing suicide in growing numbers with every passing year; the horrendous effects of pesticide sprays. The bigger picture that rarely makes news however is that millions of people are still underfed and where they do get enough to eat, the food they eat has the capability to eventually kill them.

Another negative effect of this trend has been on the fortunes of the farming communities worldwide. Despite this so-called increased productivity, farmers in practically every country around the world have seen a downturn in their fortune. The only beneficiaries of this new outlook towards food and agriculture seem to be the agro-chemical companies, seed companies and - though not related to the chemicalisation of agriculture, but equally part of the "big money syndrome" responsible for the farmers' troubles - the large, multi-national companies that trade in food, especially food grains.

Organic farming has the capability to take care of each of these problems. Besides the obvious immediate and positive effects organic or natural farming has on the environment and quality of food, it also greatly helps a farmer to become self-sufficient in his requirement for agro-inputs and reduce his costs. Thus in short, need of organic farming have been felt following reasons:

- The demand for organic food is steadily increasing both in the developed and developing countries with an annual average growth of 20-25 per cent.
- Decline in productivity of soil.
- Indiscriminate use of pesticides affects human and animal health, biodiversity, wildlife etc. & cause environmental pollution.
- High cost of inputs in conventional agriculture.
- Declining factor productivity.
- Deficiency of micronutrients.
- Global warming due to rise in carbon-dioxide and temperatures.
- Growth rate of agriculture production (1.5%) is much below the population growth rate (2.0%). Our country to be economically strong should improve on agriculture and allied enterprises.

2.1 Large Scale Farming

Today's chemical farms have little use for the skilled husbandry which was once the guiding principle of working the land. The emphasis today is solely on productivity - high input in exchange for high returns and productivity (mostly diminishing now however for farmers worldwide). Four important considerations - what happens to the land, the food it produces, the people who eat it and the communities which lose out - are overlooked. The residue persistence in agricultural produce and global fertilizer consumption as shown below in table 2.1 & 2.2 respectively.

Table 2.1: Pesticide residue persistence in agricultural produce and food commodities

Commodity	2001		2002	
	Samples (nos.)	Contamination	Samples (nos.)	Contamination
Vegetables* (17 crops)	7126	1 529 (12%above MRL#)	63.5 (8.5% above MRL)	
Fruits** (12 crops)	387	53 (Less than MRL)	329 47.0 (approaches MRL)	

* At Hisar all contaminated – 46% above MRL, Heptachlor and Cypermethrin
** Fields in Faridabad – Vegetables, fruits, and flowers highly contaminated
# Maximum residue limit (MRL)
Source: CCS Haryana Agricultural University (2003)

Table 2.2: Global Fertilizer Consumption (Mt nutrients)

N	P_2O_5	K_2O	Total
100.8	38.5	29.1	168.4
98.3	33.9	23.1	155.3
102.2	37.6	23.6	163 5
104.2	40.6	27.5	172.2
107.8	40.6	27.7	176.6
107.6	40.4	28.0	176.0

Source : IFA Agriculture, Dec 2013

Table 2.3 below shows impacts of processes on these different life and related processes on surface of the soil.

Process/methods	Impact/ Fate of processes
Land exhaustion	The constant use of artificial fertilizer, together with a lack of crop rotation, reduces the soil's fertility year by year.
Fertilizers	High yield levels are produced by applying large quantities of artificial fertilizers, instead of by maintaining the natural fertility of the soil.
Nitrate run-off	About half of the nitrate in the artificial fertilizer used on crops is dissolved by rain. The dissolved nitrate runs off the fields to contaminate water courses.
Soil erosion	Where repeated deep ploughing is used to turn over the ground, heavy rains can carry away the topsoil and leave the ground useless for cultivation.
Soil compaction	Damage to the structure of soil by compression is a serious problem in areas that are intensively farmed. Conventional tillage may involve a tractor passing over the land six or seven times, and the wheeling's can cover up to 90 per cent of a field. Even a single tractor pass can compress the surface enough to reduce the porosity of the soil by 70 per cent, increasing surface run-off and, therefore, water erosion.
Agricultural fuel	As crop yields grow, so does the amount of fuel needed to produce them. European farmers now use an average of 12 tons to 1km² of land.
Biocide sprays	The only controls used against weeds and pests are chemical ones. Most crops receive many doses of different chemicals before they are harvested.
Cruelty to animals	On most "modern" farms, all animals are crowded together indoors. Complex systems of machinery are needed to feed them, while constant medication is needed to prevent disease. The cruelty involved in managing, breeding, growing and slaughtering farm animals today is unimaginably repulsive and horrifying.
Animal slurry	With so many animals packed together in indoor pens, their manure accumulates at great speed. It is often poured into lagoons which leak into local watercourses, contaminating them with disease-causing organisms and contributing to algae-blooms.
Imported animal feed	Many farms are not self-sufficient in animal feed; instead they rely on feed brought into the farm. This often comes from countries which can ill afford to part with it.

Process/methods	Impact/ Fate of processes
Stubble burning	In countries where stubble is burned, large amounts of potentially useful organic matter disappear into the sky in clouds of polluting smoke.
Loss of cultivated biodiversity	Large and other chemical farms tend to be monocultures growing the same crop and crop variety
Threat to indigenous seeds	Native cultivars and animal breeds lose out to exotic species and hybrids. Many native animal breeds are today threatened with extinction. The same holds true for many indigenous plant varieties which have disappeared within the space of one generation.
Habitat destruction	Agribusiness farming demands that anything which stands in the way of crop production is uprooted and destroyed. The wild animals and plants which were once a common sight around farms are deprived of their natural habitat and die out.
Contaminated food	Food, both plant and animal products, leaves the farm contaminated with the chemicals that were used to produce it.
Destruction of traditional knowledge systems and traditions	Rural indigenous knowledge and traditions, both agricultural and non-agricultural, is invariably connected to agriculture and agricultural systems.
Control of agriculture inputs and food distribution Channel	The supply and trading in agricultural inputs and produce is in the hands of a few large corporations. This threatens food security, reducing the leverage and importance of the first and the last part of the supply chain - the farmer and the consumer.
Threat to individual farmers	Chemical agriculture is a threat to their livelihoods and changes their lifestyles, unfortunately not for the better.

Organic Versus Conventional

Organic farming is not a "new" concept; however, it was marginalized against the large-scale chemical based farming practices that have steadily dominated food production over the last 45 years. The difference between organic farming and modern conventional farming accounts for most of the controversy with claims and counter claims surrounding organic agriculture and organic food. Presently the comparison looks something like given below in table 2.3

2.2 Definitions

Organic farming is a reaction against large scale farming/exhaustive farming. A large number of terms are used as an alternative to organic farming. These are: biological agriculture, ecological agriculture, do-nothing agriculture, bio-dynamic, organic-biological agriculture and natural agriculture. The essential concept of these practices remains the same, i.e., back to nature, where the philosophy is to feed the soil rather than the crops to maintain soil health and it is a means of giving back to the nature what has been taken from it (Funtilana, 1990).

Table 2.3 Comparison of organic and conventional farming

Parameter	Organic farming	Conventional farming
Size	Smaller, marginal, dependent operations	Large scale, economically tied to major food corporation
Method	No use of purchased fertilizer and other inputs e.g. Pesticides, weedicides etc. less mechanization of the growing and harvesting process. Use of organic inputs like green manure, vermicompost, biofertilizers etc.	Heavy use of chemicals e.g. fertilization, use of pesticides etc. mechanized production using special equipment and facilities.
Technology	Nature based, environment friendly and sustainable.	Synthetic, harmful to environment and nutrient depleting.
Products	Good in taste, flavour, nutrition and free from chemicals.	Tasteless, less nutritious, may contain toxic residues of chemicals.
Market	Local, direct to consumer, on farm stands and farmers markets and through special wholesalers and retailer.	Wholesale with products distribution across large areas (average supermarket produce travels 100 to 1000 Km) and sold through high-volume.

According to the *Codex Alimentarius Commission,*'organic agriculture is a holistic production management system that avoids use of synthetic fertilizers, pesticides and genetically modified organisms, minimizes pollution of airs, soil and water, and optimizes the health and productivity of interdependent communities of plants, animals and people'.

Organic farming is the form of agriculture that relies on crop rotation, green manure, compost, biological pest control, and mechanical cultivation to maintain soil productivity and control pests, excluding or strictly limiting the use of synthetic fertilizers and synthetic pesticides, plant growth regulators, livestock feed additives, and genetically modified organisms. Thus, Organic Farming implies recycling of waste and residue to the native soil itself, replenishing the nutrients depleted from the soil during the crop growth, encouraging the growth of microorganisms which could regulate phased release of stored nutrients in the soil to the crop growth in right proportion, maintaining soil health by balancing the soil moisture and soil aeration and ensuring soil fertility by firmly binding the nutrient elements in the complex organic molecules.

Other Options of Organic Farming

1)Pure Organic Farming

This accounts complete exclusion of inorganic fertilizers and pesticides, but advocates the use of organic manures and biological pest control methods.

2)Integrated Green Revolution Farming

Under this option, the basic trends of the green revolution such as intensive use of external inputs, increased irrigation, development of high yielding and hybrid varieties as well as mechanizations of labour are retained with much greater efficiency on the use of these inputs with limited damage to the environment and human health. For this purpose some organic techniques are developed and combined with the high input technology in order to create Integrated Systems such as, "Integrated Nutrient Management" (INM), "Integrated Pest Management" (IPM) and biological control methods which reduce the need for chemicals.

3) Integrated Farming System

This option involves low input organic farming in which the farmers have to depend on local resources and ecological processes, recycling of agricultural wastes and crop residues.

4)Biodynamic farming

Biodynamic farming is done through the steps given below :

➢ To apply zodiac principles to the process of cosmic integration; that is to say, the moon constellations in seed sowing and harvesting of crops.

➢ To apply biodynamic methods for compost preparation and field sprays.

➢ Use of vermicompost, vermi-wash.

➢ Use of bone meal and wood ash.

➢ Use of plant-extract, cakes to manage soil and plant pests.

➢ Use of cow urine mixed with some plant chemicals, such as menthol or alone   to control plant diseases and pests.

➢ *Equisetum arvense* has been used to control fungal diseases successfully.

➢ *Melia azadirach* (Drek,) is used like Neem to control pests in the form of an extract.

➢ Plants like *Lantana camara* can be used to control insect pests or as a mosquito repellent.

➢ Use of cow pat pit or liquid manure or compost heap methods provide sufficient nutrients to the growing crops

2.3 Components of Organic Farming

There are assumptions throughout the organic literature of differences between organic and conventional systems with respect to their effects on soil physical properties, soil insect fauna and nutrient flow within the soil, crop health and nutritional value of the harvested crop. Different components of organic farming are as follows:

(a) Crop and Soil Management

Organic farming system encourages the use of rotations and measures to maintain soil fertility. Carefully managed soil with a high production of humus offer essential advantages with respect to water retention ion exchange, soil erosion and animal life in the soil. Green manuring and inter-cropping of legumes is another important aspect for biological farming systems not only in regard to weed control but also in reducing the leaching of nutrients and in reducing soil erosion. A green cover throughout most of the year is one of the main goals of such farming methods. Depending on the green manure mixture or the legumes used for under sowing, there may be an increased soil organic matter and soil N2 as well as in other nutrients.

(b) On-farm Waste Recycling

Increase price of chemical fertilizers have enables organic wastes to regain an important role in the fertilizer practices on the farm. Good manure management means improved fertilizers value of manure and slurry and less nutrient losses. Composting of all organic wastes in general and of Farm Yard Manure (FYM) or feedlot manure in particular is important in organic farming.

(c) Non-chemical Weed Management

Weed management is one of the main concernsin organic agriculture. Generally, all aspects of arable crop production play an important role in a system approach to problems. The elements to consider in preventing weed problems are crop rotation, green manuring, manure management and tillage. Mulching on a large scale by using manure spreaders may also be useful in weed control

(d) Domestic and Industrial Waste Recycling

Sewage and sludge use for crop production can form an important component of organic farming if treatment and application methods are improved further.

(e) Energy Use

In the energy requirement for production measured per rupees of produce for organic farms is only one third of what it is for their conventional counterparts. Because N-fertilizer and pesticides are not used by biological farmers, the comparison of total energy input/ha with total energy output favours biological farming systems.

(f) Food Quality

Food quality is one of the main issues, which concerns both scientists and consumers. Nitrates in water and farm produce, desirable components, pesticides residues, keeping quality and physiological imbalances are some of the important aspects of food quality.

(g) Ecological Agriculture

The growing concern about environmental degradation, dwindling natural resources and urgency to meet the food needs of the increasing population are compelling farm scientist and policy makers to seriously examine alternative to chemical agriculture. As reported by Vankataramani (1995) case studies shows that when chemical farm incurred about 11.250 towards the cost of cultivation of rice. An organic farm spend rupees 10,590 to produce 5625 kg paddy and 8 tonnes of straw/ha. The net returns from the ecological farming system at the current cost of rupees 3.34/kg paddy is rupees 8,197.50. In chemical farming, the net profit is rupees 7500. If one gets a premium price for the poison force, organically grown rice, the economic returns from the ecological farming system will highly encouraging.

(h) Integrated Intensive Farming System (IIFS)

IIFS involves intensive use of farm resources. To be ecologically sustainable, such intensification should be based on techniques which are knowledge intensive and which replace to the extent possible, market purchased chemical inputs with farm grown biological inputs.

(i) Value of Organic Farming

The value of organic materials as fertilizers and soil conditions is often misunderstood and has been the source of some controversy. The simplest and the most common means of estimating the value of organic amendments is by assigning the market value of the potentially available plant nutrients, they contain nitrogen-phosphorus-potash (NPK) and other micro-organisms.

Chapter 3

Benefits of Organic Farming

It is clear that organic agriculture, rather than confining itself to technical issues of agronomy, livestock management and the farm business, is intended to deliver much wider benefits to:

- ➢ The agricultural system
- ➢ The environment
- ➢ Society
- ➢ The economy
- ➢ Institution

The benefits of organic agriculture are expected to be environmental, social and economic as explained below.

In light of the fact that organic farms do not use synthetic products, the risk of water pollution is greatly diminished. Organically-tended soils also show reduced rates of nitrate pollution in the water supply, as organic farms use fewer nitrates than conventional farms, and organic soils have an increased capacity to retain that.

Organic farms also aim at consuming less energy and being more energy efficient than conventional farms. Studies show that they consume about forty five to sixty-four percent of the non-renewable energy (fossil fuels) consumed by conventional farms. Depending on the climate and crops studied, organic farms were found to be between twenty-five and eighty-one percent more energy-efficient.

Table 3.1 Potential Benefits of Organic Agriculture in different systems

Parameter	Potential benefits
Agriculture	Increased diversity, long term soil fertility, high food quality reduced pest/disease, self-reliant production quality stable production

Environment	Reduced pollution, reduced dependence on non-resources, negligible soil erosion, wildlife protection, protection, resilient agro ecosystem compatibility of production with environment
Social conditions	Improved health, better education, stronger community reduced rural migration, gender equality, increase employment, good quality work
Economic conditions	Stronger local economy, self-reliant economy, income security, increased returns, reduced cash investment, low risk etc
Organizational/Institutional	Cohesiveness stability, democratic organizations, enhanced capacity

The environmental benefits of organic agriculture can also extend to climate change. The International Panel on Climate Change (IPCC, 2007) has strongly advocated the adoption of sustainable cropping systems such as those used on organic farms to reduce carbon emissions. Organic methods are indeed expected to result in lower emissions – carbon emissions are between forty-eight tosixty-six percent lower than on conventional farms. This is due to the high levels of organic matter found in organic soils, which allow the soil to trap and convert carbon, lowering emissions over time.

Organic farms also tend to reduce nitrous dioxide emissions, simply because they use less nitrogen than conventional farms. This is particularly significant in light of the fact that agriculture today is responsible for sixty-five to eighty percent of nitrous dioxide pollution, which contributes to the depletion of the ozone layer. Organic agriculture is beneficial to nature protection and biodiversity conservation. Organic farmers rely on biodiversity for their success. To ensure against crop-failure, for example, organic farmers plant genetically diverse crops, thus perpetuating a diverse gene pool while also learning which seeds will be the most resilient and productive in the long term.

Organic farmers depend on wildlife for pollination, pest control and maintenance of soil fertility. The absence of synthetic pesticides provides an improved natural habitat for birds, insects and micro-organisms in the soil. As a result of such practices, studies show that bird densities, plant populations, earthworms and insect populations are much higher on organic farms than elsewhere.

Organic agriculture eschews the use of artificial synthetic pesticides, supporting the use of local species and traditional techniques of pest management. These practices are known as Organic Pest Management(OPM). OPM requires informed decision-making and careful planning. It includes: promoting populations of natural predators that contribute to controlling weeds,disease and insects; growing the most resistant varieties of crops; improving soil health to resist pathogens; growing plants in the proper seasons, which also contributes to biodiversity; using organic-approved pest-reduction and curative products, such as larvae of pest predators. These are considered effective means of controlling pests, while also promoting a healthy and diverse ecosystem.

Furthermore, organic agriculture rejects the use of genetically modified organisms or products, including plants and animals, although the possible risks posed by such products are debated widely (and in some cases such asin the EU and Tunisia, exceptions are provided for some veterinary medical products). This is because organic principles consider that the use of GMOsde-emphasizes biodiversity and is an unnatural addition to the gene pool of cultivated crops, animals and micro-organisms living on farms. As a result,the exclusion of GMOs applies to every stage of production, processing or shipping of organic products. There is the risk that GMOs may enter organic products through cross-pollination. Organic farms can thus only ensure that there has been no intentional use of GMOs in their products.

Finally, animal health and welfare is another key issue in organic agriculture. Generally speaking, organic agriculture relies on disease preventive measures while restricting the administration of veterinary drugs to livestock. Organic livestock standards further require that animals receive adequate space, fresh air and suitable shelter. They also require specific nutritional programmes using primarily organic feeds. This is a more humane and natural approach to livestock farming, which conventional agriculture does not necessarily take into consideration. There are also possible health benefits to this approach, as these techniques reduce stress in animals which is thought to prevent diseases.

(b) Social benefits of organic agriculture

Organic agriculture may have a significant social impact on rural communities. To begin with, organic farming may lead to improved employment opportunities in local communities. Organic farming often requires more manual labour to compensate for the loss of synthetic fertilizers and pesticides, and thus generates more jobs in rural communities. The amount of extra labor required varies based on the product and farm in question –figures within Europe alone have been found to vary between countries and even studies. In general, however, the labour needed to manage an organic farm is ten to twenty percent higher than on comparable conventional farms.

Organic farmers also diversify their crops and spread their planting schedules through out the year in order to maintain biodiversity and enhance the health of the soil. This creates opportunities for year-round employment,reduces turnover and may alleviate problems related to migrant labour. Crop diversification also mitigates the effects of crop failure by spreading the risk among a wider variety of crops and products. Greater job opportunities on organic farms contribute to strengthening rural communities as well, by halting exodus to urban areas for jobs.

Organic farming has the effect of strengthening local communities and supporting rural development. In order to remain competitive, farmers must adapt to local conditions by managing labour, land and resources in a way that maximises production and remains sensitive to the environment. Doing so requires constantly experimenting with new techniques and pooling local knowledge to learn best practices.

Farmers also rely on their neighbors to maintain certain standards in order to ensure the integrity of their own air, water and soil. Collaboration on these issues strengthens ties within the community, which leads to partnerships and greater organization among organic farmers. Organised groups or cooperatives can thus pool their resources, enjoy greater access to markets, and gain leverage in trade negotiations. There is some evidence that increased co-operation results in more active participation in local government and new businesses among rural communities.

Many organic farms also incorporate fair trade principles with respect to labour welfare. Through the implementation of labour rights related to organic agricultural practices, organic producers agree upon minimum social and labour standards. To that end, farmers contribute to providing laboursess with live able wages, safe and healthy working conditions and access to social services. The organic movement believes that these social requirements are important, but recognises that specific standards can be controversial and difficult to implement across numerous countries.

Consumer protection is another cornerstone of organic agriculture. Consumers prefer organic products to those made on conventional farms because they know that organic products avoid synthetic pesticides and fertilizers, are good for the environment, and are perceived to produce foods that are healthier and taste better. Strong regulatory frameworks, whereby the government verifies organic certifications, are necessary for consumers to trust the products they purchase.

Finally, organic agriculture can contribute to food security. Although the global food supply is adequate, 850 million people still go hungry. In addition, the cost of food has risen dramatically in the past decade and there is less genetic diversity in our foods due to conventional agricultural methods. Consequently,large populations are increasingly exposed to the risk of food shortage due to disease and poverty. Organic agriculture may have the potential to meet these challenges.

Considering the fact that organic methods do not require expensive chemical inputs, organic production is considered to be a more accessible means for rural farmers to become self-sufficient. Organic agriculture also improves access to food by reducing risks of disease, increasing biodiversity and productivity over the long term, and providing a means for local production and access to food.

(c) Economic benefits of organic agriculture

Organic agriculture has seen tremendous economic growth in the last decade. This has been mainly demand-driven, as consumers have become increasingly concerned with the safety of conventionally-grown foods and the ethical downfalls of industrial agriculture. Farmers, in turn, have realised that consumers are willing to pay a premium for organically-grown foods.

This is particularly attractive to farmers in developing nations, as it is expected to provide access to lucrative and emerging markets. Income constraints currently limit consumer demand mainly to the industrialised world: organic products are generally priced higher than their conventional counterparts both to cover the higher cost of production and processing and to capture unseen savings linked to issues such as environmental protection, animal welfare, and rural development. At present, North America, Japan and the European Union represent the bulk of global sales in organic products. Nevertheless, as more countries develop economically and as their populations become increasingly educated and more affluent, demand for organic products can be expected to rise.

Continued growth, however, is dependent on economic swings and food safety concerns. As a result, organic farmers must carefully plan how best to enter such markets and obtain certifications that will be recognised where they wish to sell their products. Governments have also contributed to this growth,by subsidising conversions to organic farming, as they have recognised that organic farming can help them achieve environmental, food security, and rural development goals.

Today, organic agriculture is the fastest growing food sector in the world in both land use and market size, although this fact is tempered by the fact that it was virtually non-existent until very recently. That being said, growth rates inorganic food sales have been in the range of 20–25 percent for the last ten years. In 2002, the total market value of certified organic products was estimated at US$20 billion. By 2006, that value doubled to US$ 40 billion, and is expected to reach US$70 billion by 2012. Supermarkets have also begun to supply organic foods, an indication that the movement has reached the mainstream. Organic agriculture is now practiced in approximately 120 countries throughout the world.

Organic methods can be used to produce foods and plants as well as non-traditional agricultural products. This includes non-wood forest products(NWFP), such as nuts, mushrooms, fruits, herbs, and bush meat and plant and animal products used for medicinal or cosmetic purposes. The NWFP market is estimated to be worth US$11 billion annually. The use of organic methods in woodlands and forests also promotes environmentally-friendly uses of natural resources. Organic methods are also used for fish from fish farms and honey from apiculture farms. These are both nascent markets, which have found willing consumers in developed countries.

Consumers buy organic products because they expect a certain standard of production that is environmentally-friendly and free of any artificial inputs. Organic certification ensures those standards are met and is essential for consumer-trust and expansion of the organic market. Organic consumers are at tuned to the type of certification a product receives in order to assess what quality of product they purchase. Organic certification also ensures in spection and compliance, harmonises standards across different countries and facilitates sales agreements. Certification bodies ascertain that the products meet the standards that are set by either private or public institutions. As will be seen,these standards are created either by in-country legislation or by entities at the international level.

The Sustainability Issues

Ecological Perspective

Most of the conventional farm practices are ecologically unsustainable for natural resources and soil fertility enhancing erosion and greenhouse forcing. Sustainable agricultural practices are intricately linked with ecological sustainability in terms of:

1. improved soil fertility,

2. increased ability of top soil to retain organic matter, nutrients and water,

3. increased diversity of crops, microbes and other plants and animals in and around the field,

4. reduced use of hazardous chemicals including pesticides;

5. minimised soil erosion, landslides and improved green cover to conserve soil,

6. increased carbon sequestration and,

7. reduced energy demand.

Economic sustainability

Agriculture can be sustainable only if it has a long-term economic viability. Conventional agriculture, which follows the principle of diminishing return, may pose long-term economic risks than its sustainable counterpart. The issues that need concern are:

1. **Export vs. local orientation:** From economic perspective, an export-oriented production system is considered more important than those that supply domestic demands. The Indian organic produce market is mainly export oriented. Focusing on export alone involves hidden costs including transport and risks to local food security. Policies considering domestic demands particularly food security as equally important are needed for a rationale balance of trade.

2. **Debt:** The green revolution raised India's grain production by multifold. At the same time, a large number of small-scale farmers trapped into debt. They took loans to raise production and on failure in re-paying, about 40,000 desperate farmers committed suicide.

3. **Market risk:** Concentrating on specific commodities although promises high economic returns, is vulnerable to market risks. Market fluctuates quickly and a disproportional sweep of low priced international agricultural produce into the national market, may lead Indian farmers at risk. As a WTO signatory, the Indian government is under pressure to open its economy to the global market and hence unable to protect farmer's interest especially those of small farm holders.

4. **Employment:** Agriculture is the main source of employment for rural people.

Specialised and mechanised practices reduce rural employment. Sustainable agriculture, as witnessed through organic farming system, being labour-intensive helps overcome such problems.

Social sustainability

The social sustainability of farming techniques focuses on social acceptability and justice. Ignoring these issues may lead to loss of valuable local knowledge and provoke political unrest. This needs serious attention as our country has a large share of traditional knowledge and culture. The issues that need serious concern include:

1. **Inclusiveness**: Development is a process of organized and dimensional growth driven by negative feed backs. It cannot be sustainable unless it is inclusive reducing poverty for the broad masses of people. This has particular concern for countries like ours having very large gap between rich and poor. The government needs to explore ways to enable rural poor to get benefit from agricultural development.

2. **Political unrest**: Rising gaps between rich and poor feed social injustice driving poor masses to feel neglected and excluded from developmental opportunities. The result is the political unrest, violence, and economic instability.

3. **Local acceptance**: Many new technologies fail for being based on practices followed outside. Sustainable agricultural practices consider local social customs,traditions, norms and taboos. Thus, the local acceptance enhances harmony, fulfill needs and promote sustained growth and yield.

4. **Indigenous knowledge:** India is among the leading countries regarding its climatic, biotic and cultural diversities. Our country has a vast treasure of tribal diversity and traditional knowledge. Sustainable agriculture often focuses on the use of traditional knowledge and local innovation. Locally adapted breeds and crop varieties coupled with their social structures to manage and conserve common resources, can support strengthen stability in agriculture. A balance use of indigenous knowledge with appropriate information added from outside would drive sustainable agricultural to enrich itself.

5. **Gender:** In our traditional agriculture, women bear the heaviest load in terms of labour. In modern conventional farming men often benefit the most by controlling what to grow and how to spend the resulting income. Sustainable agriculture ensures that the loads and benefits to be shared more equitably between men and women.

6. **Food security**: Modern farming approaches in India consider few crops only and fail to provide variety and a balanced diet. Sustainable agriculture ensures food security by improving the quality and nutritional value of food with greater range of crop varieties and edible produce.

7. **Participation:** Traditional society in India is driven by wealth and caste distinctions. Conventional farming innovations often exacerbate these gaps. Sustainable agriculture consciously targets the less well-off people as well. From

social point of view, sustainable agriculture involves full participation of vibrant rural communities to ensure safe and sustained food supply for everyone.

There is no real dispute that sustainable agriculture and organic farming are closely related terms. There is however disagreement on the exact nature of this relationship. For some, the two are synonymous, for others, equating them is misleading. Lampkin's definition of organic farming, quoted above, talks of sustainable production systems. Having provided his definition, he goes on to state:-sustainability lies at the heart of organic farming and is one of the major factors determining the acceptability or otherwise of specific production practices" (Lamp-kin, 1994, p. 5). Similarly, Henning *et al.,* (1991) precede their definition of organic farming, quoted above, by claiming that "it could serve equally well as a definition of 'sustainable agriculture'" Rodale even suggested that "sustainable was just a polite word for organic farming". Despite the variety of definitions of organic farming, the general agreements regarding what is necessary to produce organically are in stark contrast to the debates and arguments that rage regarding the nature of agricultural sustainability. However, as Ikerd (1993) notes, "mention 'sustainable agriculture' and many people will think you are talking about organic farming. Some organic farmers will agree.

They think that organic farming is the only system that can sustain agricultural production over the long run" (p. 30). This view of an extremely close if not synonymous relationship between organic farming and sustainability is not universal, and it should of course be noted that the elusive nature of sustainability's definition and meaning imply that equating it to anything is a rather bold step. Hodge argues against those like Bowler (1992), who view organic farming as the only truly sustainable type of agriculture, contending that this is only true if non-sustainability is identified through the use of non-renewable resources, especially inorganic chemicals.

In opposition to this position he states that: " it must be questionable as to whether organic farming, as currently practised, can reasonably be regarded as sustainable" (Hodge, 1993, p. 4). Factors that Hodge uses to support his argument include uncertainty regarding nitrate losses from conventional and organic farming, particularly in light of the difficulty in controlling nutrient applications from organic manures. Concerns over the long-term maintenance of potassium levels in soils, especially on dairy farms, and the issue of soil erosion are also cited.

The conclusion drawn is that "it is thus a mistake to equate 'sustainable' agricultural systems with 'organic' ones. A restriction on the use of inorganic chemicals is not a sufficient condition for sustainability, but it may not even be a necessary condition" (Hodge, 1993, p. 4). Pretty (1995, p. 9) argues that although "organic agriculture is generally a form of sustainable agriculture", it can also have negative environmental effects.

These include the leaching of nitrates from field under legumes, the volatilisation of 26 D. Rigby, D. CaÂceres / Agricultural Systems 68 (2001) 21±40 ammonia from livestock waste and the accumulation of heavy metals in soil following the application of Bordeaux mixture. Some of the research that has been carried out regarding the historical relationship between agricultural systems and the sustainability of the societies they support, illustrates the point that a farming system need not be modern, mechanised, and using synthetic chemicals to be profoundly unsustainable. Carter and Dale (1974), in a historical review of the relationship between the soil, agricultural systems and the civilizations they have supported, explain how the fertility of large areas of Greece, Lebanon, Crete and North Africa was destroyed by low input, chemical-free unsustainable agricultural practices. The farmers whose agricultural practices contributed to this erosion and desolation were undoubtedly organic producers in terms of the inputs used, but they were 'organic by neglect'. This point is not merely of historical interest, examples of the organic by neglect approach are still witnessed today. Hall, an organic inspector with the Organic Crop Improvement Association1 (OCIA) in the USA, states that this idea that a crop is organic because 'nothing has been put on it' is all too common.

This, he argues, is not a sustainable approach and "does a major disservice to the majority of organic farmers who are making excellent progress in developing healthy and naturally resilient whole farm systems" (Hall, 1996).These points support the view that focusing on particular inputs or tools in the identification of sustainable agricultural systems is insufficient.

In response it might be argued that inputs and tillage methods are only one part of the picture that organic production goes beyond these narrow production issues. Lampkin and Measures (1995, p. 3) write that "the term 'sustainable' is used in its widest sense, to encompass not just conservation of non-renewable resources (soil, energy, minerals) but also issues of environmental, economic and social sustainability." The IFOAM aims refer to the need "to interact in a constructive and life-enhancing way with natural systems and cycles to consider the wider social and ecological impact of the organic production and processing system to encourage and enhance biological cycles within the farming system, involving micro-organisms, soil flora and fauna, plants and animals to progress towards an entire production, processing and distribution chain which is both socially just and ecologically responsible" (IFOAM, 1998, p. 3). Clearly, the standards do not exist in a vacuum they represent an attempt to move from general principles, such as these from IFOAM, to specific practices and inputs, whether recommended or prohibited.

The difficulty is that incorporating these wider concerns into definitions of, and standards for, organic farming is problematical. Standards are far more able to refer to prohibited inputs than to deal with precise criteria for the assessment of whether producers and processors are acting in a manner which is "socially just" or "ecologically responsible". The significance of this increases when one considers the massive expansion of the organic sector currently underway in many countries, where the motivations of newly converting organic producers may well be different from the 'traditional' organic producer who associated closely with these broader principles.

This issue of the range of motives that people may have for adopting organic techniques must be carefully considered. While many adopting organic practices are doing so for lifestyle and more holistic reasons, the issue of higher market prices for organic goods cannot be ignored. Lampkin and Measures (1995) report, for example, organic prices in the UK between 50 and 100% above conventional prices for cereals and vegetables. It seems highly probable that these economic factors are driving the conversion decision for many new organic producers in contrast to the past. In the UK, this changing profile of the new organic producer is a result of the number of established conventional producers who are now converting to organic production, something which the number of calls to the Organic Conversion Information Service (OCIS) reflects.

In the context of the prolonged crisis in large sections of British agriculture the possibility must therefore exist of producers becoming organic to pursue these premiums; their motive may not be sustainability in its broadest sense, but marketing at its most strategic. A greater understanding of the range of motives for adopting organic techniques is needed, and the implications of this range of motives for any discussion of the relationship between organic and sustainable farming practices must be considered.

Weymes (1990) found that 9% of the Canadian organic farmers surveyed stated that profitability was their primary reason for adopting organic farming (Rigby *et al.*, 2000). Fairweather and Campbell (1996) found that over a third of the organic farmers they interviewed would switch to conventional production if premiums decreased, and on the basis of an analysis of these organic producers distinguished between "pragmatic organic" and "committed organic" farmers. Part of the difficulty here is that these organic schemes must focus on prohibiting or encouraging the use of particular inputs or tools, whereas it is the use of these things that determines a system's sustainability. Stolze *et al.* (2000) argue that organic farming uses two methods to obtain environmental results: "the regulation of the use of inputs" and "the requirement of specific measures to be applied or, in some cases, of the outcome of environmental or resource use".

The authors confirm the emphasis on the regulation of inputs explaining that "the first method is more important and the second is more a supplement". This orientation on specific inputs is hardly surprising since these schemes require producers to either be registered or not; there can be no grey areas, the produce is sold either with the organic symbol, or without. The criteria must therefore be clear, well-defined and open to inspection. Objectives such as the sustainability of farm families, farm workers and rural communities, which are frequently espoused by organic groups, are simply not amenable to this type of regulation. Individual producers may be committed to such goals, but most standards do not include them, and it is difficult to see how they could.

The issues discussed above point towards a rejection of the view that organic farming is simply the practical implementation of sustainable agriculture's principles, or indeed that, as has been claimed, it represents the pinnacle of sustainable agriculture. This does not imply that organic agriculture is unsustainable. Rather, the notion of sustainability is such a "site-specific, individualistic, dynamic concept" (Ikerd, 1993), that arguing that one particular set of codified production practices are its practical expression seems incorrect and likely to attract unnecessary criticism. In this sense, the sustainability concept may be viewed similarly to appropriate technology, in that the appropriateness of particular technologies will also vary temporally and spatially (McInerney, 1978).

The information that is required to inform this debate further is detailed data regarding the environmental impacts of organic production systems. Such information is sparse, although the increased interest in the sector over recent years has produced a series of initiatives investigating these matters, some of which have reported. Stolze *et al.* (2000) provide a review of the environmental impacts of organic farming in Europe based on a review of existing literature in national and international sources noting that "as data availability on the subject has not always been satisfying, a qualitative analysis has been chosen as an approach" (2000).

The impacts are assessed under seven headings: ecosystem; soil; ground and surface water; climate and air; farm input and output; animal health and welfare and quality of food produced. Summarizing some of their findings, the authors find that "organic farming clearly performs better than conventional farming in respect to oral and faunal diversity" however "direct measures for wildlife and biotype conservation depend on the individual activities of the farmers" (Stolze *et al.*, 2000). In terms of soil it is concluded that "organic farming, tends to conserve soil fertility and system stability better than conventional farming systems .

No differences between the farming D. Rigby, D. CaÂceres / Agricultural Systems 68 (2001) 21±40 29 systems were identified as far as soil structure is concerned". Regarding water quality the review concluded that "organic farming results in lower or similar nitrate leaching rates than integrated or conventional agriculture". Conclusions regarding the impacts on climate and air are hard to draw because of a lack of data and the difference between calculations per unit of land as opposed to per unit of output. Stolze *et al.* (2000) conclude that nutrient balances on organic farms are often close to zero and that "energy efficiency' is found to be higher in organic farming than in conventional farming in most cases" .Work on impact assessment raises the issue of which are the key aspects of a system's performance that should be measured, that is, what are the key aspects of agricultural sustainability and what are the associated indicators that should be monitored. Stolze *et al.* (2000) adapt the OECD set of environmental indicators, using only those indicators which directly affect the system of organic farming.

This issue of indicator development is a rapidly developing area of work which is reviewed by Glen and Pannell (1998); Moxey (1998) and Rigby *et al.* (1999). Specific examples of work on constructing indicators of agricultural sustainability are to be found in Taylor *et al.* (1993); Gomez et al. (1996); Swete-Kelly (1996); Bockstaller *et al.* (1997); MuÈller (1998) and Rigby *et al.* (2000). Part of the difficulty in assessing the sustainability of agricultural systems, an issue which many of the papers cited above address, is the fact that both the units of measurement and the appropriate scales for measurement differ both within and across the commonly identified economic, biophysical and social dimensions of sustainability.

For example, consideration of the effects of organic production on farm margins, soil fertility and rural employment are difficult to combine in an overall measure. Not so problematic if the effects are all in the same direction, but when one starts to consider trade-offs, as one indicator increases and another falls, across different dimensions then this factor becomes more significant. This is an issue which will not be solved simply by greater knowledge of the impacts of different production systems; even with complete information regarding impacts one will still have to consider trade-offs with movement towards targets in some respects accompanied by reverses in others. Despite this complication of trade-offs and the need for judgments to be made about priorities, the notion of sustainability as a goal, a signpost rather than a destination, is still useful (Ikerd, 1997). Thought of in this way, the convergence to agricultural sustainability may be viewed as an asymptotic process.

Cost-benefit Issues

Agriculture is the base of economic policies and is the ultimate driver of national economic growth and poverty alleviation in many developing countries including India. The industrial agriculture however, that increased grain production and farmers profit by a large margin, is being driven by significant externalities with long standing hidden cost such as loss of natural resources, effects on human health and on agriculture itself (Subba Rao, 1999). Organic farming has now been tagged not only for minimizing externalities but also for its cost effectiveness.

Model estimates indicate that organic methods have potential to produce enough food to sustain current human population and an even a larger population without increasing the agricultural land area while reducing the detrimental effects of conventional agriculture (Badgley *et al.*, 2007). Some government programmes in Sweden, Canada, and Indonesia have demonstrated that organic farming can reduce pesticide use by 50% to 65% without sacrificing crop yields and quality along with 50% lower expenditure on fertilizer and energy use (Pimentel *et al.*, 2005).

The increasing demand for organic produce has created new export opportunities and many developing countries have started to tap lucrative export markets for organic produce. Export of tropical fruits and African cotton to the European food industry, Zimbabwe herbs to South Africa, and Chinese tea to the Netherlands and soybeans to Japan are classical examples of organic food export market.

Indian organic farming industry is almost entirely export oriented, running as contract farming under financial agreement with contracting firms. Further, the majority of farmers in India are opting this practice motivated by attractive market and price margins (Sharma, 2001). Thus, the capital driven policies coupled with lack of open local market for sale of organic produce may negatively influence the bottom-up response on organic farming discouraging small farm holders who have currently no access to organic agricultural technology and certification. Cost-benefit analysis (CBA), sometimes called benefit-cost analysis (BCA), is an economic decision-making approach used particularly in government and business sectors.

It compares the total expected costs of each option against the total expected benefits, to asses if the benefits outweigh the costs and with what margin.

The benefit cost- ratio (BCR), the ratio of net value of crop produce (minus cost of inputs) to cost of input that depicts total financial return for each rupee invested in this production system (IGNOU, 2007), is an important tool to assess economics of farming. The production system is considered viable if the ratio is more than one. For agriculture sector, the component of cost estimate includes fixed costs, variable cost and other costs. Fixed cost includes land, land revenue, depreciation of farm implements and interest on fixed capital. Variable cost includes cost of planting materials, organic inputs, pesticides, irrigation, bullock, tractor and cost of labour and irrigation and other costs include cost of marketing, power consumption, storage and packing.

A study, based on 120 farmers of six villages of Shimoga and Bhadravati Talukas of Karnataka State of India, compared the cost-benefit components of organic rice production (Suresh and Kunnal, 2004). The study indicated that in organic farm, although the average cost of cultivation per acre of paddy was lower only marginally, the net return increased by over 40% suggesting that a properly planed organic farming is beneficial not only from environmental point of view but also from economic margin.

Another study undertaken by Central Institute for Cotton Research, Nagpur India indicated that the cost of cultivation under organic farm was about 21 % lower than that under conventional farm mainly due to no use of chemical fertilizers and insecticides(IGNOU, 2007). An increase in price margin subject to market demand of organic produce status further substantiates total benefits. Similarly, the benefit cost ratios estimated for per acre organic and inorganic wheat and carrot were found to be 1: 1.08; 1: 1.01 and 1: 1.52; 1: 1.44 respectively (Mehmood *et al.*, 2011). In particular, for rain-fed systems organic agriculture out-performs its conventional counterpart (Ramesh *et al.*, 2005). A survey of 208 projects in developing countries, in which contemporary organic practices have been introduced, revealed average yield increases of only 5–10% in irrigated crops and 50–100% in rainfed crops (Ptetty and Hine, 2001).

In another study, it was observed that, despite reduction in crop productivity by 9.2%, organic agricultural produce provided a 22.0% higher net profit to farmers due to coupled effect of 11.7% reduction in cost of production and 20-40% greater premium price of certified organic produce (Ramesh *et al.*, 2010). The successive improvement in soil quality in organic farming constitutes an important hidden benefit as it helps reducing cost of future fertilizer needs (Escobar and Hue, 2007). Production costs of organic and conventional farming is given in Table 3.2 (Sharma, 2013).

Table 3.2: Production Costs of Organic and Conventional Farming

Sr.No.	Operation/Inputs	Conventional	Organic
1	Land Preparation	2400	2400
2	Seed	1400	300
3	Seed inoculants	0	50
4	Sowing	500	500
5	Intercultural operations	700	400
6	Vermicompost/FYM	0	1000
7	Trichocompost	0	250
8	Basal Fertilizers	2220	0
9	Urea	400	0
10	Amrut pani (a fermented mixture	0	0
	of cowdung and cattle urine)		1000
11	DAP spray	125	0
12	Plant protection chemicals	0	0
13	Biological control agents	0	0
14	Trichcards	0	0
15	Chrysoperla spp.	0	0
16	Ha NPV	0	0
17	Harvesting	2000	1800
	TOTAL	**1118**	**8850**

(Source: Sharma, PD, 2003)

Table 3.3 Organic Comsumtion Periods

Year	Status	Yield (Qtls/ha)	Premium 20%	Total (Rs)	Net Income (Rs)	Surplus/Deficit Over conventional cotton
Conventional		10.0	2000	0	900	0
First Year	Under conversion	5.00	10000	0	750	-8250
Second Year	Under conversion	5.75	11250	0	3750	-5250
Third Year	Organic	6.25	12500	2500	7000	-1500
Fourth Year	Organic	7.5	15000	3000	1050	1500
Fifth Year	Organic	8.75	17500	3500	13500	4500
Sixth Year	Organic	10.00	20000	4000	16500	7500

(Source: Sharma, P.D. 2003)

Chapter 4

Principles of Organic Farming

The Four Principles of Organic Agriculture are as follows:

4.1 Principle of Health

Organic Agriculture should sustain and enhance the health of soil, plant, animal, human and planet as one and indivisible.

This principle points out that the health of individuals and communities cannot be separated from the health of ecosystems - healthy soils produce healthy crops that foster the health of animals and people.

The role of organic agriculture, whether in farming, processing, distribution, or consumption, is to sustain and enhance the health of ecosystems and organisms from the smallest in the soil to human beings. In particular, organic agriculture is intended to produce high quality, nutritious food that contributes to preventive health care and well-being. In view of this it should avoid the use of fertilizers, pesticides, animal drugs and food additives that may have adverse health effects.

4.2 Principle of Ecology

Organic Agriculture should be based on living ecological systems and cycles, work with them, emulate them and help sustain them.

This principle roots organic agriculture within living ecological systems. It states that production is to be based on ecological processes, and recycling. Nourishment and well-being are achieved through the ecology of the specific production environment. For example, in the case of crops this is the living soil; for animals it is the farm ecosystem; for fish and marine organisms, the aquatic environment.

Organic farming, pastoral and wild harvest systems should fit the cycles and ecological balances in nature. These cycles are universal but their operation is site-specific. Organic management must be adapted to local conditions, ecology, culture and scale. Inputs should be reduced by reuse, recycling and efficient management of materials and energy in order to maintain and improve environmental quality and conserve resources.

4.3 Principle of Fairness

Organic Agriculture should build on relationships that ensure fairness with regard to the common environment and life opportunities.

This principle emphasises that those involved in organic agriculture should conduct human relationships in a manner that ensures fairness at all levels and to all parties - farmers, workers, processors, distributors, traders and consumers. Organic agriculture should provide everyone involved with a good quality of life, and contribute to food sovereignty and reduction of poverty. It aims to produce a sufficient supply of good quality food and other products. This principle insists that animals should be provided with the conditions and opportunities of life that accord with their physiology, natural behaviour and well-being.

Natural and environmental resources that are used for production and consumption should be managed in a way that is socially and ecologically just and should be held in trust for future generations. Fairness requires systems of production, distribution and trade that are open and equitable and account for real environmental and social costs.

4.4 Principle of Care

Organic Agriculture should be managed in a precautionary and responsible manner to protect the health and well-being of current and future generations and the environment.

Organic agriculture is a living and dynamic system that responds to internal and external demands and conditions. Practitioners of organic agriculture can enhance efficiency and increase productivity, but this should not be at the risk of jeopardizing health and well-being. Consequently, new technologies need to be assessed and existing methods reviewed. Given the complete understanding of ecosystems and agriculture, care must be taken.

This principle states that precaution and responsibility are the key concerns in management, development and technology choices in organic agriculture. Science is necessary to ensure that organic agriculture is healthy, safe and ecologically sound. However, scientific knowledge alone is not sufficient. Practical experience, accumulated wisdom and traditional and indigenous knowledge offer valid solutions, tested by time. Organic agriculture should prevent significant risks by adopting appropriate technologies and rejecting unpredictable ones, such as genetic engineering. Decisions should reflect the values and needs of all who might be affected, through transparent and participatory processes.

Chapter 5

Different Sources /Practices Used in Organic Farming

5.1 Composting

A huge quantity of crop and animal wastes/residues is always available on a farm. The common practice is to burn plant wastes, which, besides being an environmental disaster, is also a waste of the huge potential of these residues.

Fig. 5.1 showing Compost as organic manure

Properly recycled, these residues form excellent compost in one to six months, depending upon the composting process used. Every farm can choose or even develop a suitable compost process depending upon its own needs and resources, including availability of labour, managerial time and investment potential, soil health, as a pest repellent and prophylactic, in composting, and in animal feeds, animal health and hygiene, aquaculture, etc. Following figure 5.1 shows Compost as organic manure

Table 5.1: Average nutrient composition of organic material (oven-dry basis)

Kind of material	Total-N	Total P_2O_5	Total K_2O	Total CaO
Compost	1.34	3.30	1.04	0.89
Swine manure	0.81	3.00	0.61	4.75
Carabao manure	0.60	2.05	0.50	-
Cow manure	1.87	2.47	2.11	'-
Goat manure	2.81	2.66	1.20	-
Horse manure	3.13	2.80	1.88	-
Sludge	1.87	3.11	0.54	4.30
Vermi-cast	1.86	3.61	1.60	2.21
Azolla	2.76	0.97	2.38	1.09
Rice straw	0.48	0.34	1.58	-
Coconut coir dust	0.50	0.82	1.26	-
Mud press	2.72	6.20	0.79	-
Distillery slops	0.12	0.25	0.62	-
Garbage ash	0.68	T	1.40	3.45
Water hyacinth ash	0.50	8.06	19.08	-
Factory ash	0.22	2.76	0.94	0.75
Bagasse ash	0.28	0.84	2.00	-

5.2 Vermicomposting

Vermicomposting is a modified and specialised method of composting - the process uses earthworms to eat and digest farm wastes and turn out high quality compost in two months or less. Vermicompost is not a biofertiliser as is touted by some, merely improved compost. Vermicompost can also be used to make compost tea. Vermicompost tea is useful as a prophylactic against pests and diseases, for pest repelling and as a foliar spray. A by-product of vermicomposting called vermi-wash also serves the same purpose. An important point to note in case of vermicomposting but widely ignored, is to carry out proper sieving of the compost before applying it in the fields. The most efficient and widely used earthworms in vermicomposting are not indigenous and if the worms and casts find their way to the fields, they will quickly colonise and dominate the local species. Farmers can also use indigenous earthworm species, collecting them from their fields using collection baits and introducing the earthworms into heaps.

In the usual way vermicomposting is practiced in India and most other places around the world, it is both labour-intensive and requires some infrastructure. As a result, while a small farm can use this method to compost most of its wastes, a larger farm often finds it expensive and difficult to compost most of wastes through vermicomposting. Table 5.1shows the average nutrient content of different organic

Fig. 5.2 Versatile waste eater and decomposer *Eisinia foetida*

wastes and Vermicompost and other Composts (Table 5.2) and table 5.3 shows average secondary & micro nutrient status of vermicompost and Farmyard manure respectively while figure 5.2 shows Versatile waste eater and decomposer *Eisinia foetida*.

Table 5.2:   Average nutrient content of Vermicompost and other Composts

Compost Nutrient Content (% of dry matter)	*N*	*P_2O_5*	*K_2O*
Vermicompost	1.6	2.2	0.7
Rural compost	1.2	1.1	1.5
Urban compost	1.2	1.9	1.5
Paddy straw compost	0.9	2.1	0.9
Maize stalk compost	1.1	1.3	1.0
Cotton wastes compost	1.6	1.1	1.5
Water hyacinth	2.0	1.0	2.3
Poultry manure	2.9	2.9	2.4
Castor	5.8	1.8	1.0
Cotton seed	3.9	1.8	1.6
Neem	5.2	1.0	1.4
Niger	4.8	1.8	1.3
Rapeseed	5.1	1.8	1.0
Linseed	5.5	1.4	1.2
Sunflower	4.8	1.4	1.2

Table 5.3: Average secondary and micro-nutrient contents of vermicompost and FYM

Nutrients	Vermicompost	FYM
Ca (%)	0.44	0.91
Mg (%)	0.15	0.91
Fe (ppm)	175.2	146.5
Mn (ppm)	96.51	69.0
Zn (ppm)	24.43	14.5
Cu (ppm)	4.89	2.6

(Source: Fertilizer News, 2004)

Suhane (1982) studied the chemical and biological properties of soil under organic farming (using various types of composts) and chemical farming (using chemical fertilizers-urea (N), phosphates (P) and potash (K)). Results are given in Table 5.4 below:-

Table 5.4: Farm soil properties under organic farming and chemical farming

Chemical and biological properties of soil	Organic farming (Use of composts)	Chemical farming (Use of chemical fertilizers)
1) Availability of nitrogen (kg/ha)	256.0	185.0
2) Availability of phosphorus (kg/ha)	50.5	28.50
3) Availability of potash (kg/ha)	489.5	426.5
4) Azotobacter (1000/gm of soil)	11.7	0.80
5) Phospho bacteria (100,000/kg of soil)	8.80	3.20
6) Carbonic biomass (mg/kg of soil)	273.0	217.0

(Source: Suhane, (2007)

5.3 Enriched City Compost

City compost produced at mechanical composting plants throughout the country is generally low in plant nutrients and therefore its acceptability by farmers has been limited. To improve the quality and nutrient content of city compost low-grade rock phosphate and phosphate solubilising *Azotobacter* spp. and the nitrogen fixing bacteria, such as *Azotobactor* spp. or *Pseudomonas* spp. are being used as inoculants. Microbial inoculation and application of 1 to 5 percent rock phosphate increased the nitrogen content of city compost by 24 to 30 percent and more favourable C: N ratios have been obtained. Available P_2O_5 content of compost was increased by 60 to 114 per cent where rock phosphate was applied and inoculated with *Aspergillus awamori*.

Preparation of compost from enriched city garbage or otherwise is promising, provided that financial support from government is available. However, heavy metals in sewage sludge when continuously applied in excessive quantities to farmland as organic manure could lead to problems. Monitoring for Cd, Zn, Pb. As, and Cu contents in compost is recommended. A comparison of chemical compost with ordinary compost is given in table 9.

Table 5.5: Showing Chemical Composition of Phospho-Compost Compared With Ordinary Compost

Component	Ordinary-compost	Phospho-compost	Rock phosphate
Total N (%)	1.3	0.95	-
pH	8.6	8.8	8.8
Total P (%)	0.25	3.00	8.4
Organic P (%)	0.05	0.06	0.18
Available P (mg/g)	1.20	0.63	0.18
Water-solubleP (mg/g)	0.31	0.20	-
Citric acid-soluble (mg/g)	0.52	12.24	2.20

5.4 Bio-fertilizers

In strict sense, are not fertilizers, which directly give nutrition to crop plants. These are cultures of microorganisms like bacteria, fungi, packed in a carrier material. Thus, the critical input in biofertilizer is the microorganisms. They help the plants indirectly through better Nitrogen (N) fixation or improving the nutrient availability in the soil. The term "Biofertilizer" or more appropriately a "Microbial inoculants" can generally be defined as preparation containing live or latent cells of efficient strains of Nitrogen fixing, Phosphate solublising or cellulolytic microorganisms used for application to seeds, soil or composting areas with the objective of increasing the number of such microorganisms and accelerate those microbial process which augment the availability of nutrients that can be easily assimilated by plants. Biofertilizer can provide an economically viable support to small and marginal farmers for realizing the ultimate goal of increasing productivity. Biofertilizer are low cost, effective and renewable source of plant nutrients to supplement chemical fertilizers. These organisms are added to the rhizosphere of the plant to enhance their activity in the soil. Sustainable crop production depends much on good soil health. Soil health maintenance warrants optimum combination of organic and inorganic components of the soil. In nature, there are a number of useful soil microorganisms that can help plants to absorb nutrients.

Their utility can be enhanced with human intervention by selecting efficient organisms, culturing them and adding them to soils directly or through seeds. Biofertilizers are living microorganisms of bacterial, fungal and algae origin. Their mode of action differs and can be applied alone or in combination. By systematic research, efficient strains are identified to suit to given soil and climatic conditions. Such strains have to be mass multiplied in laboratory and distributed to farmers. They are packed in carrier materials like peat, lignite powder in such a way that they will have sufficient shelf life. The biofertilizer are mainly purchased by State Agriculture Departments and distributed to the farmers at concessional rates. About 200 to 500 grams of carrier material is only needed per acre, costing about Rs.10/- to 25/.

Bio-fertilizers, or bio-inoculants comprise environment friendly microorganisms which are beneficial to agriculture to improve soil fertility or crop productivity. They supply nutrients (ex. nitrogen) as well as improve availability of the unavailable forms of certain others (ex. phosphorus) and are comprised by several bacteria, fungi, actinomycetes etc. *Rhizobia, Azotobacter, Azospirillum,* blue green algae, *Azolla* and phosphate solubilizers (several bacteria and fungi) are the key examples. Their role in supplementing nutrition makes them ideally suitable in integrated nutrient management systems. The interactions among the rhizosphere, the roots of higher plants and the soil borne microorganisms have a significant role in plant growth and development. The organic compounds, released by roots (Lévai, 2004) and bacteria (Katznelson and Bose, 1959), play an important role in the uptake of mineral nutrient. In recent years, bio-fertilizers have emerged as an important component of the integrated nutrient supply system and hold a great promise to improve crop yield through better environmentally nutrient supplies. Table 5.6 below shows list of Bio fertilizers, their Functions and Beneficiaries Crops.

Fig. 5.3 Shows *Azolla* in Paddy

Table 5.6: List of Biofertilizers, their Functions and Beneficiaries Crops

S. No	Name of the biofertilizer	Function / Contribution	Beneficiaries(Crops)
A.	**Nitrogen Biofertilizers**		
1.	Rhizobium(symbiotic)	Fixes 50-100 kg N/ha2. Increase yield from 10-35%3. Leaves residual nitrogen	Pulse legumes, Oilseed legumes, Fodder egumes, Forest legume
2.	Azotobacter (non symbiotic)	Fixation of 20-25 kg N/ha10-15% increase in yield Production of growth promoting substances	Wheat, Maize,Cotton, Sorghum, Sugarcane, Pear, Millet, Rice,Vegetables and Several other Crops
3.	Azospirillum(associative)	Fixation of 20-25 kg N/ha2.10-15% increase in yield.3. Production of growth promoting substances	Wheat, maize, Cotton, sorghum,Sugarcane, pear, Millet, rice,Vegetables and Several otherCrops
4.	Blue green algae(BGA)(phototropic)	Fixation of 20-30 kg N/ha10-15% increase in yield. Production of growth promoting substances	Flooded rice
5.	Azolla(symbiotic)	Fixation of 30-100 kg N/ha2. 10-15% increase in yield	Only for flooded rice
B.	**Phosphorus Biofertilizers**		
1.	PhosphateSolubilizingMicroorganis ms(Bacteria/fungi)	Solubilizers of insoluble phosphates. Yield increase from 10-20%	All types of crops
2.	VAM(Obligate symbionts)	Enhance uptake of P, Zn, S, Fe, Cu and water, Promotes uniform crop, increase growth	Forest trees

(a)*Rhizobium* Inoculants

The nitrogen fixed by rhizobia benefits legume crop production in two ways:(a) by meeting most of the legume crops nitrogen needs and (b) by enriching the soil for a the benefit of subsequent crops. *Rhizobium* inoculation should be considered in al legume green manure crops to gain maximum benefit from nitrogen fixation in the shortest possible time .*Azospirillum, azotobacter* and *pseudomonas* inoculations on upland rain crops are still in their infancy and field trial results are inconclusive, although good responses to *azospirillum* and *azotobacter* inoculation of wheat, rice, sugarcane and mustard have been recorded.The contribution of rhizobium in biological nitrogen fixation in various crops is given table 5.7 below.

Table 5.7: Contribution of Rhizobium in Biological Nitrogen Fixation

S. No.	Name of the crop N fixation	Quantum of (kg/N/ha)
1.	Alfalfa	100-200
2.	Redgram	166-200
3.	Clover	100-200
4.	Gram	85-110
5.	Cowpea	80-85
6.	Groundnut	50-60
7.	Lentil	90-100
8.	Green/Black gram	50-55
9.	Pea	52-77
10.	Soybean	60-80

The figure 5.4 below represents the diagrammatic nodule formation in legume and life cycle of the Rhizobium bacteria.

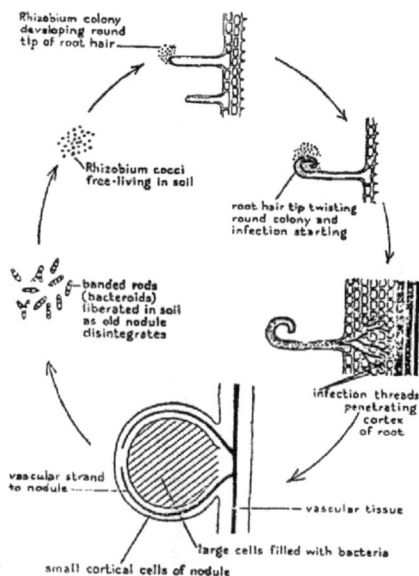

Fig. 5.4 Diagrammtic Representation of Nodule Pormation in L legume, and the Life Cycle of the Root nodule Backeria (*Rhizobium* sp.)

Source: Coble, Leslie S., An Introduction to the Botany of Tropical Crops. New York: Longman, Green and Company

Effect of bio-fertilizers (N-fixers) on rice yield– A case study

The aim of this study was to evaluate the efficiency of bio-fertilizers in an integrated manner on few ruling varieties of rice for low land (ADT37, IET1444, and CO47) and upland (Chinna ponni, ADT37 and White ponni) under Pondicherry conditions in three different seasons of a year I.e. spring-summer (April-July), Autumn-winter (August-Jan) and winter-spring (Jan-April) (Figure 5.5. and 3).

Effect of biogertilizers on rice yield in lowland system, T_1: Contro; T_2 100% N Inorg. Ferti; T_3: 75% n Inorg. Ferti; T_4: T_3+ *Azospirillum*; T_5: T_3+ BGA; T_6; T_3 *Azospirillum* + BGA Sabhashini et al. 2007

The results indicated that under lowland system, in the initial stage (spring-summer), the inorganic fertilizers (100% =T_1) resulted in the highest yield (2088 kg/acre in ADT37), compared to others. A slight change was observed in the second crop, where T3 showed the highest yield followed by T_2 and T_4, with no significant difference between them. However, in winter-spring season the yield of CO47 in T_6 was on par with T_2 and T_3 (1200, 1231 and 1139 kg/ acre. It shows a gradual increase in efficiency of biological and its compatibility with inorganic fertilizers.

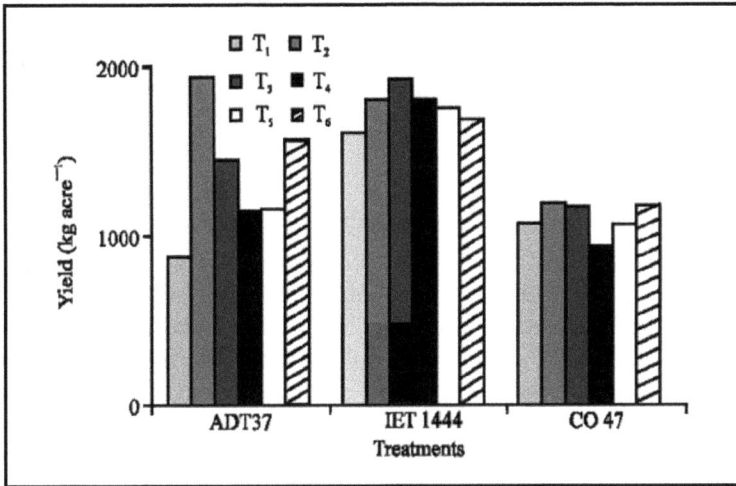

Figure 5.5

Effect of bio-fertilizers on upland rice yield (Figure 3)

The results clearly show the positive correlation between the usage of microbes and consistent increase in crop yield. The highest yield during first season (winter-spring-chinna ponni) was observed in T2 (1289kg/acre), while in two subsequent generations, T6 recorded the highest yield (1705 and 1400 kg/acre respectively) followed by T4 and T5, confirming the efficiency of bio-fertilizers and compatibility with chemical fertilizers with time. These results have also showed the enhancement of N, P and K after post-harvest stage indicates the steady increase in the fertility status and water holding capacity of soil due to residual effect.

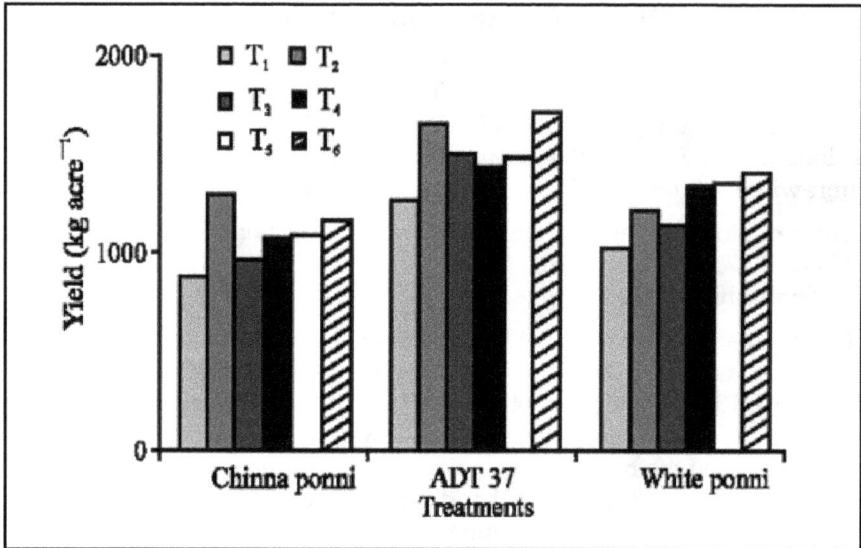

Fig. 5.6

Effect of biogertilizers on rice yield in lowland system, T_1: Contro; T_2 100% N Inorg. Ferti; T_3: 75% n Inorg. Ferti; T_4: T_3+ A*zospirillum*; T_5: T_3+ BGA; T_6; T_3 A*zospirillum* + BGA Sabhashini et al. 2007

Sabhashini *et al.* 2007

(b) Phosphate Solubilizing Microorganisms (PSM)

A variety of bacteria and fungi has been identified to have the ability to solubilise and transform inorganic P from normally insoluble sources through excretion of various organic acids have been isolated. These are bacteria of the *bacillus* and *pseudomonas* app and fungi, such as aspergillus, penicillium and trichoderma spp. In addition to p-solubilisation these microorganisms can also mineralise locked up organic P into soluble, plant available forms. As these reactions take place in the rhizosphere and the microorganisms bring more P into solution than they can absorb for their own growth, the surplus is available for plants to absorb. The effectiveness of these microorganisms depends on the availability of sufficient energy source. Carbon in the soil, P concentration, particle size of rock phosphate as well as temperature and moisture.

Mechanisms of Phosphorus Solubilization

Some bacterial species have mineralisation and solubilisation potential for organic and inorganic phosphorus, respectively (Hilda and Fraga, 2000; Khiari and Parent, 2005). Phosphorus solubilising activity is determined by the ability of microbes to release metabolites such as organic acids, which through their hydroxyl and carboxyl groups chelate the cation bound to phosphate, the latter being converted to soluble forms (Sagoe *et al.*, 1998). Phosphate solubilisation takes place through various microbial processes / mechanisms including organic acid production and proton extrusion (Surange, 1995; Dutton and Evans, 1996; Nahas, 1996).

General sketch of P solubilisation in soil is shown in Figure 5.7 A wide range of microbial P solubilisation mechanisms exist in nature and much of the global cycling of insoluble organic and inorganic soil phosphates is attributed to bacteria and fungi (Banik and Dey, 1982). Phosphorus solubilisation is carried out by a large number of saprophytic bacteria and fungi acting on sparingly soluble soil phosphates, mainly by chelation-mediated mechanisms (Whitelaw, 2000). Inorganic P is solubilised by the action of organic and inorganic acids secreted by PSB in which hydroxyl and carboxyl groups of acids chelate cations (Al, Fe, Ca) and decrease the pH in basic soils (Kpomblekou and Tabatabai, 1994). The PSB dissolve the soil P through production of low molecular weight organic acids mainly gluconic and keto gluconic acids (Goldstein, 1995; Deubel *et al.*, 2000), in addition to lowering the pH of rhizosphere. The pH of rhizosphere is lowered through biotical production of proton / bicarbonate release (anion/cation balance) and gaseous (O_2/CO_2) exchanges. Phosphorus solubilisation ability of PSB has direct correlation with pH of the medium.

Release of root exudates such as organic ligands can also alter the concentration of P in the soilsolution (Hinsinger, 2001). Organic acids produced by PSB solubilize insoluble phosphates by lowering the pH, chelation of cations and competing with phosphate for adsorption sites in the soil (Nahas, 1996). Inorganic acids *e.g.* hydrochloric acid can also solubilise phosphate but they are less effective compared to organic acids at the same pH (Kim *et al.*, 1997). Figure 4 shows the schematic diagram of soil phosphorus mobilisation and immobilisation by bacteria.

(c) Mycorrhiza

The term mycorrhiza denotes "fungus roots". It is a symbiotic association between host plants and certain group of fungi at the root system, in which the fungal partner is benefited by obtaining its carbon requirements from the photosynthates of the host and the host in turn is benefited by obtaining the much needed nutrients especially phosphorus, potassium, ammonium ions, zinc, copper etc., which are otherwise inaccessible to it, with the help of the fine absorbing hyphae of the fungus. These fungi are associated with majority of agricultural crops, except with those crops / plants belonging to families of *Chenopodiaceae, Amaranthaceae, Caryophyllaceae, Polygonaceae, Brassicaceae, Commelinaceae, Juncaceae and Cyperaceae*. They are ubiquitous in geographic distribution occurring with plants growing in arti\c, temperate and tropical regions .The fungi that are probably most abundant in agricultural soils are arbuscular mycorrhizal (AM) fungi. They account for 5–50% of the biomass of soil microbes (Olsson *et al.*, 1999). Biomass of hyphae of AM fungi may amount to 54–900 kg ha-1 (Zhu and Miller, 2003). Pools of organic carbon such as glomalin produced by AM fungi may even exceed soil microbial biomass by a factor of 10–20. The mineral acquisition from soil is considered to be the primary role of mycorrhizae, but they play various other roles as well which are of utmost important.

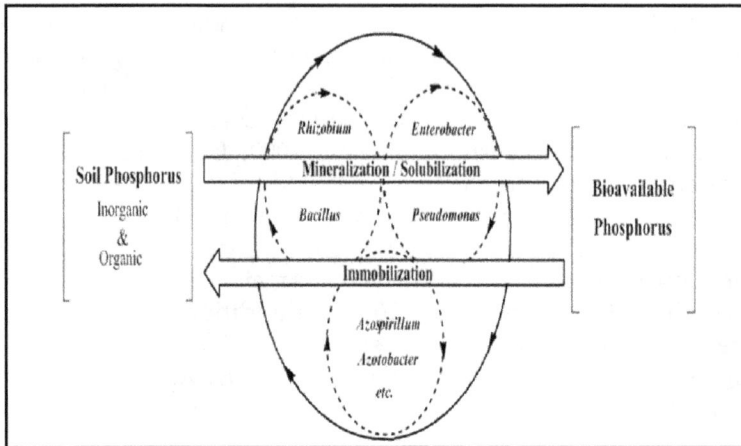

Fig. 5.7: Schematic diagram of soil phosphorus mobilization and immobilization by bacteria.

Table 5.8 Response of vegetable crops to *Rhizobium, Azospirillum* and *Azotobacter* inoculation

Biofertiliser	Crop	Increase in yield (%)	Nitrogen economy (%)
Rhizobium	Pea	13.38	-
Azotobacter	Cabbage	8.60	25
	Garlic	14.23	25
	Knol knol	9.60	25
	Onion	12.16	25
Azospirillum	Cabbage	10.58	25
	Capsicum	9.98	25
	Kolkhol	14.90	25
	Onion	21.68	25
	Onion	7.74	25
	Garlic	6.42	25

Source :Marwaha (1995)

5.5 Mulching, Green manuring and Cover cropping

All these techniques are different but somewhat interrelated.

Mulching is the use of organic materials (plastic mulch is expensive and non-biodegradable) to cover the soil, especially around plants to keep down evaporation and water loss, besides adding valuable nutrients to the soil as they decompose. Mulching is a regular process and does require some labour and plenty of organic material, but has excellent effects, including encouraging the growth of soil fauna such as earthworms, preventing soil erosion to some extent and weed control.

Plate 5.8 Sesbania and Crotalaria as green manure crops

Green manuring is an age-old practice prevalent since ancient times. A crop like *dhaincha (Sesbania aculeata)*, sun hemp or horsebean is sown (usually) just before the monsoons. A mix is also possible. Just around flowering (30-45 days after sowing), the crop is cut down and mixed into the soil after which the season's main crop is sown. Green manuring is beneficial in two ways - firstly it fixes nitrogen, and secondly the addition of biomass (around five to ten tons/acre) greatly helps in improving the soil texture and water holding capacity. Green leaf manuring can also be carried out if sufficient leguminous tree leaves are available. The table 5.9, given below shows the percent N fixing capacity and productivity (T/Ha) of different green manures.

Fig. 5. 9 Cowpea and forage grass as cover crops

Table 5.8 Bio fertilizers as a component of IPNS in the different agro eco-regions.

Ecosystem	Ecological Region	Location	Crop/Cropping system	Soil	Bio-fertilizers as IPNS Component
Arid	Western Plain, Kachchh and part of Kathiawar Peninsula, Hot arid ecoregion	Hissar(Haryana)	Pearlmillet	Sandy loam	FYM @ 2.5 tonnesha-1 + 20 kg Nha-1 + seed treatment with biofertilizer
			Wheat		Recommended dose of 120 kg N + 60 kg P2O5 + 60 kg K2Oha-1 + Inoculation of wheat seeds with azotobacter
Semi-arid	Northern plain (central highlands) including Aravallis, Hot semi-arid	Udaipur (Rajashthan)	Mustard	Sandy loam	75% NPKS recommended dose based on soil test + 10t FYMha-1 75% NPKS based on soil test + Azotobacter + PSB
		Junagadh (Gujarat)	Chickpea	Medium black	20 kg N + 40 kg P2O5ha-1 + Rhizobium inoculation
		Amreli (Gujarat)	Sesame	Vertisols	50% N through urea + 50% N through FYM + PSB (50% P, 100% K)
	Deccan plateau, hot semi arid	Parbhani (Maharashtra)	Summer groundnut	Clay	50% NPK through chemical fertilizer + 50% NPK through FYM + Azotobactor + cowdung urine slurry + phosphate solubilizing bacteria
			Okra	Medium black clay with slightly alkaline	75 kg N + Biofertiliser (Azotobactor) + FYM @ 10 t ha-1
			Greengram Blackgram Groundnut	Vertisols	Rhizobium inoculatrin + Molybdenum @ 4 g / kg of seed treatments

Contd....

Ecosystem	Ecological Region	Location system	Crop/ Cropping	Soil	Bio-fertilizers as IPNS Component
Semi-arid	Deccan Plateau	Pune(Maharashtra)	Cotton-sorghum	Black soils	FYM @ 6.2 ha^{-1} Deep tillage in sorghum80 kg BGA ha^{-1}
		Secunderabad (Andhra Pradesh)	Rice Bengalgram Groundnut	Sandy loam	100% NPK + Rhizobium 2 kg ha^{-1} + Phosphobacteria 2 kg ha^{-1}
	(Telangana) and eastern ghat, Hot semi arid		Sunflower		P-enriched FYM + Azospirillum+ phosphobacteria + 0.2% Borax
	Eastern Ghat and Tamil Nadu uplands and Deccan (Karnataka) plateau, hot semi-arid	Maruteru (A.P.)	Black-gram	Vertisols	VAM inoculation + 50 kg P_2O_5ha^{-1}
		Coimbatore (Tamilnadu)	Bajra	Clay loam	Azospirillum + 100% NP + phosphobacterium
			Rabi rice	Sandy clay loam	GM (*Sesbania aculeata* @) 6.25 tha^{-1}) or FYM (12.5 tha^{-1}) with Azospirillum (2 kgha^{-1}) and N 150 kgha^{-1} with 25 kg $ZnSO_4$ha^{-1}
		Vriddachalam	Groundnut	Red sandy	FYM + Rhizobium + phosphobacterium
		Coimbatore (Tamilnadu)	Blackgram	Black soils	75% P_2O_5 as Tunisia rock phosphate + vermicompost @ 2 kg ha^{-1} + phosphobacteria @ 2 kgha^{-1}
Semi-arid		Pudukkottai (Tamil Nadu)	Blackgram	Acidic soil	Biodigested slurry @ 5 tha^{-1} + Rhizobium
		Pudukkottai	Greengram	Acidic soil	Compost @ 5 tha^{-1} + rhizobium
		Paiyur (Tamil Nadu)	Fingermillet -horsegram	Loamy sand	Recommended inorganic fertilizers + FYM 750 kgha^{-1} or 75% recommended fertilizer dose (RFD)+ biofertilizers

Contd...

Table 5.8: Contd...

Ecosystem	Ecological Region	Location	Crop/Cropping system	Soil	Bio-fertilizers as IPNS Component
		Paiyur (Tamil Nadu)	Sorghum -horsegram	Loamy sand	Recommended inorganic fertilizers + FYM @ 10 tha⁻¹ or 75% recommended inorganic fertilizer + biofertilisers
Sub-humid	Central Highlands (Malwa and Bundelkhand), hot sub-humid (dry)	Sehore (M.P.)	Pigeonpea and blackgram	Vertisols	FYM sugar pressmud @ 5 t⁻¹ with rhizobium inoculation
			Soybean	Vertisols	FYM @ 6 t ha⁻¹ or sugar press mud @ 5 t⁻¹ with rhizobium inoculation
		Bhawanipatna (Odhisha)	Cotton	Loamy clay	N, P and K @ 100, 50 and 50 kgha⁻¹ in conjunction with Azotobacter and PSM @ 5 kgha⁻¹ each and FYM @ 10tha⁻¹.
	Western Himalaya Warm sub- humid to humid	Palampur (H.P.)	Potato	Clay loam	75% of recommended inorganics + 20 tha⁻¹ FYM

Table 5.9: shows the percent N fixing capacity and productivity (T/Ha) of different green manures

Crop	Productivity (T/ Ha)	Nitrogen%
Subabul	09-11	0.80
Sunhemp	12-13	0.43
Dhaincha	20-22	0.43
Cowpea	15-16	0.49
Clusterbean	20-22	0.34
Berseem	15-16	0.43

(Source: Maity and Triparthy, 2005)

These are one-year-cycle crops, cultivated before tree plantation or in the alley during orchard lifetime with the aim, other than the addition of OM to the soil, of fixing and trapping nutrients, reducing nitrate leaching, nutrient run off and soil erosion, improving soil structure and aeration, reducing nematodes infestations. Suitable cover crops are legumes species such as clover, broad bean, pea which fix atmospheric N at a rate of 20-200 kg N/ha (Adjei *et al.*, 2008; Rahman *et al.*, 2004), especially if soil shows a poor concentration of mineral N. Non-leguminous species, such as cereals (i.e. oat, barley, rye-grass, sorghum) and *Brassica* species (rape, mustard, etc.) are preferred when nutrient availability is high in the fall so that they can trap P and N that otherwise would be lost by leaching or runoff. Most *Brassica* species such as *B.napus, B. nigra, B. alba* etc., are biocide plants and reduce infestation of some fungi (i.e. *Rhizoctonia* spp.) and nematodes (i.e. *Pratylencus penetrans*) due to the production of isothiocyanates, volatile compounds from the digestion of vacuole located glucosinolates (Brown *et al.*, 2008) by myrosinase enzyme. A mix of grasses and legumes may be preferred to combine the tolerance to low winter temperature of non-leguminous species with the positive effect on N_2 fixation of legumes. Cover crops can be sowed in autumn, late winter or early spring according to the temperature requirements, and usually incorporated into the soil early in summer, when crops are at the beginning of flowering stage and the C:N ratio is lower than 20:1.

5.6 Crop Rotation and Polyculture

One of the most Important aspects of organic farming is the strict avoidance of monoculture, whether annuals or perennials. Besides the proverbial "putting all eggs into one basket", monoculture systems are unhealthy for the ecosystem they are a part of. The prime requirement for any natural ecosystem to thrive and be healthy is diversity. Traditional farmers till date follow the systems of crop rotation, multi-cropping, intercropping and polyculture to make maximum use of all inputs available to them, including soil, water and light, at a minimum cost to the environment. The home gardens of Kerala are an excellent example.

Crop rotation

Crop rotation is the sequence of cropping where two dissimilar type of crops follow each other - a few examples include cereals and legumes, deep-rooted and short rooted plants and where the second crop can make use of the manuring or irrigation provided some months earlier to the first crop (eg. rice + wheat, rice + cotton). The combinations possible are endless, and will depend to a great deal on the local situations.An example of crop rotation is below:

Rotation 1	A 4 year potato / mixed vegetable rotation
Year 1	Red clover – cut and mulch
Year 2	Potatoes – grazing rye winter green manure
Year 3	*Brassicas*
Year 4	Mixed vegetables + winter green manure as appropriate

The inclusion of main crop potatoes and, potentially, a wide range of *brassicas* mean that this rotation requires relatively large amounts of land. Composted manure or compost should be applied where appropriate but at an equivalent rate of no higher than 25t/ha (10t/acre). The positions of the *brassicas* and mixed vegetables could be reversed providing sufficient compost is applied to what would be the last block in the sequence. Crop rotation of brown sarson + rice is the most beautiful example of two crop rotation in temperate conditions of Kashmir valley becoming popular day by day among the farming community, fits well on the rice –oilseed rotation because brown sarson can withstand abnormal weather conditions viz. frosty winter and prolonged showers in spring with very low temperatures as shown in plate 6.

Plate 5.10: Brown sarson (*rabi*) followed by rice (*Kharif*) in Kashmir(J & K).

Multi-cropping

Multi-cropping is the simultaneous cultivation of two or more crops. In Indian agricultural tradition, farmers have been known to sow as many as 15 types of crops at one time. An example of multi-cropping is Tomatoes + Onions + Marigold (where the marigolds repel some of tomato's pests).

Inter-cropping

Inter-cropping is the cultivation of another crop in the spaces available between the main crop. A good example is the multi-tier system of coconut + banana + pineapple/ginger/leguminous fodder/medicinal or aromatic plants. While ensuring biodiversity within a farm, inter-cropping also allows for maximum use of resources.

Plate 5.11 and 5.12 show intercropping system

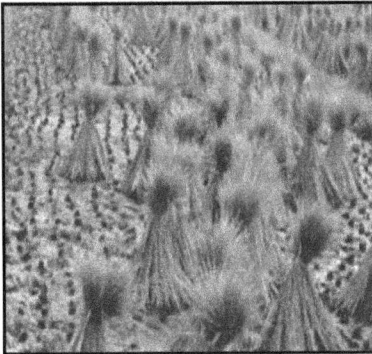
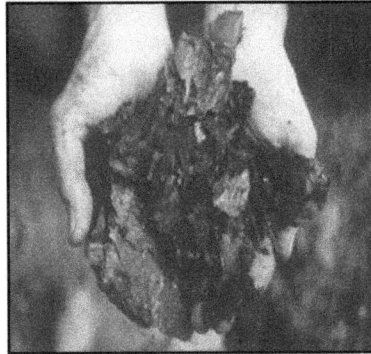

Plate 5.13: Shows crop residues and leaf litter as organic sources

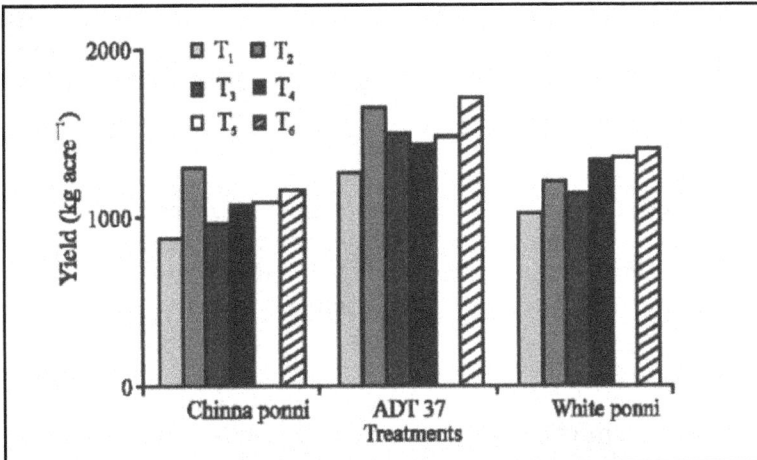

Plate 5.14: shows *In situ* moisture conservation (Fall / Summer ploughing)

pulations in control. Fallen leaves and other crop residues in combination add more value to the soil or compost heap they become a part of the nutritional resources. Plates 5.11 and 5.12 show intercropping system.

5.7 Effective Microorganisms

As the name suggests, it makes use of microorganisms, mainly lactic acid bacteria, photosynthetic bacteria, yeast, filamentous fungi and ray fungi. These microorganisms are both aerobic and anaerobic and are not genetically modified. EM, like Biodynamics can be useful in many different ways on the farm, including improving soil health, as a pest repellent and prophylactic, in composting, and in animal feeds, animal health and hygiene, aquaculture, etc.

5.8 Integration of systems

In nature, the whole is greater than the sum of its parts and the key to the success of any natural system is diversity. Diversity adds complexity to the farm system lending it greater stability. There are economic and productivity benefits too. The concept of polyculture should not be limited to plants only but extended to cover the whole farm. This way, one system's wastes and by-products are another system's inputs, or one system is comprised of more than one component, which allows for efficient use of available resources. An example of such integration is: rice-fish/prawn systems and annual crops + tree s+ cows + honeybees.

5.9 Biopesticides

Biopesticides, comprising living organisms or natural products derived from them are exemplified by plants (ex. pyrethrum *Chrysanthemum* sp., neem *Azadirachta* or *Melia* sp. etc.), macrobials (ex. *Trichogramma* parasitoid a protozoan, *Cryptolaemus montrouzieri* a coccinellid predator etc.), microscopic animals (ex. nematodes), microorganisms including bacteria (ex. *Bacillus thuriogenisis*), viruses (ex. nucleopolyhedrosis virus), fungi (ex. *Beauveria* sp.) and the transgenic plants containing a pest combating gene (ex. Bt cotton).Their key advantages include safety to mammals and other non-target organisms, environment compatibility, target specificity, lower exposure to pests, supplemental role to chemical pesticides enabling their use in integrated pest management and acceptability for use in organic agriculture. Some examples include, Bt (*Bacillus thuringiensis*) formulations to control mainly lepidopterae species. *Trichoderma* sp mainly to control soil fungi, *Verticillium lechanii* to control whiteflies, aphids, thrips and other insects, *Bauveria bassiana* to control thrips and other insects *Paecelomices fumosoroseus* controlling different insects, mainly beetles, fire ants and nematodes, *Corynebacterium paurometabolum* to control nematodes and other pests. *Metarhizium anisopliae* tocontrol termites, various coleoptera insects, leafhoppers and aphids.Similarly, biofertilizers, or bioinocculants comprise environment friendly microorganisms which are beneficial to agriculture to improve soil fertility or crop productivity. They supply nutrients (ex. nitrogen) as well as improve availability of the unavailable forms of certain others (ex. phosphorus) and are comprised by several bacteria, fungi, actinomycetes etc. *Rhizobia, Azotobacter, Azospirillum,* blue green algae,*Azolla* and phosphate solubilizers (several bacteria and fungi) are the key examples. Their role in supplementing nutrition makes them ideally suitable in integrated nutrient management systems and in organic farming. The following Table 5.10 shows different biopesticides their target pests and specific crop.

Table 5.10 shows different biopesticides their target pests and specific crop.

Biological control	Target	Crop
Bacillus thurigiensis Lepidoptera, citrus, Mites strains LBT-1, LBT-13, grass lands. LBT-21, LBT-24	Lepidoptera, Mites	Vegetables, roots and tuber, tobacco, potato, plantain,
Beauveria bassiana strainLBB-1	Coleoptera (weevils), ants, Trips palmi	Sugarcane, plantain,citrus, rice, potato,beans
Verticillium lecanii strain Y-57	Bemisia tabaci Myzus persicae	Vegetables, roots and fruits
*Metarhizium anisopliae*strain LBM-11	Lepidoptera andColeoptera	grass lands, rice and plantain
Trichoderma harzianum *Trichoderma* spp.	Phytophthora,Rhizoctonia, Phytium,Sclerotium	obacco, vegetables, T ornamentals, grains
Trichogramma pretiosum,T. pinto, *Trichogramma* spp.	Lepidoptera	Sugarcane, tobacco, Cassava, cabbage, cucumber
Phytoseiulus macropilis	Mites	Cassava, banana,plantain
Telenomus spp.	Spodoptera frugiperda	Corn, sugarcane
Pheidole megacephala	Cylas formicarius	Corn, sugarcane
Encarcia spp.	Bemisia spp.	Beans

5.10 Bio-gas Slurry as Manure

The dung and the farm wastages are increasing being burned instead of being returned to the soil as manure. Technology is available for the conversion of the dung to fuel and at the sametime retain fertilizer value of the material. The gas produced from cow dung and water as a result of anaerobic fermentation is called bio-gas. Bio-gas contains methane gas (50-65%) as most useful component and the remaining part mostly being CO_2 with small amount of other gas (Khatri-Chhetri, 1991).

Agronomic Importance of Bio-gas Slurry

A field experimental was conducted by Kuppuswamy et al. (1993) to study the effect of bio-gas slurry and gypsum riched bio-gas slurry on rice-blackgram. Bio-gas slurry at 10 tonnes/ha.enriched with gypsum @250kg/ha gave an additional grain yield of 1.8 tonnes/ha.compare to control. The residual effect of FYM on succeeding blackgram was comparatively better than that of biomass slurry.

5.11 Permanent Grass

Most of the advantages of soil management with permanent grass are those observed for cover crop and deal with addition of OM to the soil, fixing and trapping nutrients, reducing nitrate leaching, nutrient run off and soil erosion, increasing availability in the upper soil profile of nutrients with a low mobility such as K, P, and Mg (Schliemann et al., 1983); in addition, the association of fruit trees with grass species can alleviate the symptoms of Fe deficiency-induced leaf chlorosis. While the amount of organic material returned every year by the above ground mass is relatively easy to evaluate, grass root turnover and OM from rhizodeposition are often underestimated. For example, Balesdent and Balabane (1996) reported that, although the estimated aboveground (345 g C m^{-2} year^{-1}) was higher than the below-ground (152 g C m^{-2} year^{-1}) corn residue, the latter contributed more to the soil OM than did the above-ground residues vs. 36 g C m^{-2} year^{-1}). Also, the average macronutrient rates yearly returned by permanent grass were higher than those of cover crops (Table 5.11). Five years of grass soil management reduced yearly peach yield (26.3 t/ha) compared to the cover crop (31.2 t/ha) and tillage (34 t/ha), but had the positive effect of reducing nitrate N soil concentration. In addition, the presence of alley grasses reduced the incidence of downy mildew and powdery mildew in grape (Marangoni *et al.*, 2001).

Table 5.11: Direct and Residual Effect of Bio-digested slurry on Rice and Blackgram

Treatments	Rice yield (ton/ha.)	Blackgram (kg/ha)
Wet bio-digested slurry @10 ton/ha.	7.46	422
Dried bio-digested slurry @ 10 ton/ha	7.80	393
Wet bio-digested slurry @ 10 ton/ha with gypsum 250 kg/ha	8.41	402
FYM @ 10 ton/ha	7.33	463
FYM @ 10 ton/ha with gypsum 250 kg/ha	8.00	431
Gypsum 250 kg/ha	6.78	294
Control	6.61	292
CD (P = 0.05)	0.19	51

Source: Kuppuswamy et al., 1993

The release of nutrients from grass litter after mowing follows different rates according to the mineral: K is released almost completely after 5 weeks while N, Ca, and P take some months (Tagliavini *et al.*, 2007). The use of mowed grass as mulch in organic farming systems is thus not always recommended in pome fruit, because they suffer from post-harvest disorders (i.e. bitter pit) related to K:Ca ratio imbalance. All the strategies to prevent soil undesirable nutrient conditions and the consequent excessive accumulation of K in the fruit must be adopted, considering that in organic farming, fertilizers with a prompt release of Ca are unavailable. Iron management of fruit trees is a major issue in calcareous soil, where prevention of leaf chlorosis should be achieved by appropriate agronomic techniques that include: introduction of resistant or tolerant rootstocks, increase in OM soil content. The orchard management with grasses such as *Festuca* spp. may reduce or prevent Fe chlorosis through the production of phytosiderophores, natural compounds such as the mugineic acid (Klair *et al.*, 1995) that are able to chelate soil insoluble Fe.

Table 5.12 Macro and micronutrients yearly supplied by above ground cover crops and grass floor management (Giovannini *et al.*, 2001).

Macronutrients (kg/ha)					
	N	*P*	*K*	*Ca*	*Mg*
Cover crop	85	16	139	87	43
Grass	134	19	158	83	42
Micronutrients (kg/ha)					
	Fe	*Mn*	*Zn*	*Cu*	*B*
Cover crop	14	0.7	0.2	0.1	0.2
Grass	7.8	0.8	0.3	0.2	0.1

Cover crop consisting of barley (80%) and winter vetch (20%) was planted in fall and tilled in spring.Permanent grass composition included *Lolium perenne* (22%), *L. italicum* (16%), *Bromus inermis* (12%), *Festuca arundinacea* (10%), *Phleum pratensis* (10%), *Lupolina virgo*(5%), *Lupinella gusciata* (10%), and clover (10%).

5.12 Use of Agro-Industry Wastes

Press mud, coir pith, sea weed residues, cotton wastes, bagasse, biogas slurry, mushroom spent waste etc contribute substantial quantities of NPK besides secondary and micro nutrients.

5.13 Oil Cakes and Other Organic Manures

Oil cakes of non-edible types like castor, neem and karanji (*Pongamia pinnata*) as well as edible cakes like groundnut, mustard are widely used in India as organic manures due to their high NPK content. Nimbin and Nimbicidin is said to inhibit nitrification processes. Animal wastes like bone meal, fish meal etc are also rich in nutrients and are often used in organic farming. Tapping and proper utilisation of such locally available organic resources could provide substantial quantity of crop nutrients in organic farming.

5.14 Naturally Occurring Mineral Amendments

According per Codex Alimentarius Commission (COA), some naturally occurring mineral amendments are allowed in a restricted manner in organic farming to supplement the crop nutrient requirements. These include rock phosphate, rock potassium, rock sulphate, guano, basic slag, gypsum (calcium sulphate), Epsom salt (magnesium sulphate), calcite lime, dolomite lime (Singh and Dabas, 2012).Projection on the availability on the organic resources in India during 2000-25 is given in table 5.13.

Diseases in Organic Versus Conventional Agriculture

Diseases that plague conventional farming operations, causing yield loss or the application of costly inputs, are often the same species that challenge organic growers producing the same crops. One significant difference is that organic growers avoid the use of broad-spectrum synthetic pesticides, which severely disrupt natural controls in the system and promote the occurrence of secondary pests. Natural pest and pathogen controls are not only conserved (not disrupted) but are also promoted in organic farming conditions. Most soil borne plant pathogens causing root and foot rots in older plants are usually less prevalent in organic than in conventional farms. This kind of disease suppression has frequently been associated with higher microbial activity and diversity, with higher microfaunal numbers and diversity, and/or with lower soil and crop N concentrations in organic than in conventional soils. Damping-off causing pathogens such as *Pythium* species can wreak havoc in organic crops, since these can multiply quickly in fresh organic materials incorporated into soil.

Table 5.13 Projection on the Availability of the Organic Resources in India during 2000-25

Organic sources	2000	2010	2025
Human population (million)	1,000	1,120	1,300
Livestock population (million)	498	537	596
Foodgrain production (m t)	230	264	315
Nutrients (theoretical potential)			
Human excreta (million t N+ P_2O_5+K_2O)	2.00	2.24	2.60
Livestock dung (million t N+ P_2O_5+K_2O)	6.64	7.00	7.54
Crop residues (million t N+ P_2O_5+K_2O)	6.21	7.10	20.27
Nutrient (considered tapable)			
Human excreta (million t N+ P_2O_5+K_2O)	1.60	1.80	2.10
Livestock dung (million t N+ P_2O_5+K_2O)	2.00	2.10	2.26
Crop residues (million t N+ P_2O_5+K_2O)	2.05	2.34	3.39
Total	**5.65**	**6.24**	**7.75**

Source: Tiwari (2005)

Table 5.14 On-farm and field-experiment comparisons of disease levels of some important crops under organic or ecological versus conventional management

Crop	Management practices in organic crops	Consequence as compared to conventional	References
Apple	Organic soil amendment that promote soil microbial diversity	Lower colonization by root pathogens (*Rhizoctonia, pythium*)	Manici *et al.*, 2003
Almond	A mixed cover crop, no fertilizers and pesticides	Increased microbial communities	Teviotdale and Hendrics 2003
Cereals	Organic practices(no fertilizers and pesticides)	Reduce incidence and severity of root rots	Van Brugen, 2003
Potato	Absence of fungicides or only copper fungicides	Reduced *verticilium dahlia*	Lazarovits, 2001
Tomato	Biological insecticides, and living mulch cover crops and composted manure	Lower severity of corky root rot, pythium and phytopthora root rot	Clark *et al.*, 1998
Strawberry	Organic, Mulch, no insecticides and no fumigation	Cylindrocarpon root rot lower in organic	Rosado-May *et al.* 2005
Grapes	Cover crops	No difference in Phylloxera but reduced severity of fungal root rots	Clark *et al.*, 1998

Table 5.13 lists recent examples of field comparisons between organic and conventional agriculture in different crops and location. These studies show that biodiversity is generally higher on organic farms, those pests and pathogens are usually regulated by organic practices, but that there are exceptions in either case. Showing Disease Resistance in Cauliflower Induced by Vermicompost.

A). Cauliflower grown on chemical fertilizers (Susceptible to diseases)

B). Cauliflower grown on vermicompost (Resistant to diseases)

Plate 5.15: Photo Showing Disease Resistance in Cauliflower Induced By Vermicompost

Chapter 6

Organic Crop Production Technique

Response of Some Vegetables to Organic Farming

Potato: The long-term field experiment for seven years at Jalandhar (Sharma *et al*, 1988) revealed that FYM was more effective in increasing tuber yield than green manuring with dhaincha. Grewal and Jaiswal (1990) reported that the yield increase due to increased nutrients by increasing organic matter. From studies in different places, it was found that FYM to supply 100 kg P2O5.ha (about 30t/ha) not only met P and K needs of the crop but also kept the potato yield level at a higher than the combined use of P and K fertilizers (Sud and Grewal, 1990). Role of green manures in economising P and K for potato has been evaluated in the field experiments at Jalandhar (Sharma *et al*, 1988; Sharma and Sharma, 1990).

> **Tomato:** Application of oil cakes of margosa, castor, and groundnut (@0.2% W/W) is generally found to reduce the intensity of root gall development. Thamburaj (1994)found that organically grown plants were taller with more number of branches. They yielded 28.18 t/ha, which was at par with the recommended dose of FYM and NPK(120:100:100 kg/ha).

> **Brinjal**: Highest yield of brinjal was with 50 kg N/ha as poultry manure and 50 kg N/ha in the form of urea (Jose *et al*, 1988). By application of neem cake higher yield was obtained in brinjal (Som *et al*, 1992)

> **Okra:** Okra responded to poultry manure @ 20 kg N/ha (Abusaleha and Shanmugavelu, 1989). There was increased in protein and mineral content of okra crop by application of FYM as compared to commercial manures (Bhadoria *et al*, 2002). Higher yield was also recorded by application of neem cake (Raj and Geetha Kumari, 2001). Application of bio-fertilizers with chemical fertilizers increases the availability of NPK in soil and fruit in okra (Subhiah, 1991).

> **Cauliflower:** Singh and Mishra (1975) obtained highest returns of cauliflower bymulching with mango leaves.

> **Cabbage**: Application of animal compost (cattle manures and chicken manure) to mineral soil of cabbage crop was effective in reducing the

leaching out of mineral nutrients. The total carbon content was increased with the application of compost prepared with cattle manure. Nitrate content in the soil water increased with the amount of chemical fertilizers applied but remained low when only compost were applied (Nishiwaki and Noue, 1996).

Nutrient Management and Fruit Quality

Quality of fruits in present day scenario of competitive market, consumer rights and people awareness is great of importance. Organic fruits like apple, almonds, Cashew nuts, Walnuts etc., are selling at a high rate as compared to fruits grown under conventional farming. The effect of organic fertilization on quality and especially on nutritional value of fruits is still debated. In comparison to integrated or conventional systems, organic fruit management was found to increase the content of phenols (flavanols) and nutritional fibers in apple (Weibel *et. al.*, 2000), ±-tocopherol in pears, ascorbic and citric acids in peaches, but at the same time to decrease ±-tocopherol in peaches (Carbonaro *et al.*, 2002). In general an improvement of secondary metabolism products such as organic acids and polyphenolic compounds, many of which considered beneficial for human health, occurs as a consequence of the organic cultivation practices (Winter and Davis, 2006). There are two major hypotheses as a possible explanation for this response: the first involves the defense system of the plants, which promote the production of plant polyphenolics as defense mechanisms in replacement of chemical insecticides and fungicides its use is limited in organic cultivation system (Asami *et al.*, 2003). The other hypothesis is related to the nutritional status of the plant and in particular the N availability: if N is a limiting factor for the tree and consequently the ratio C: N increases, tree metabolism will move toward a higher synthesis of secondary metabolism compounds such as phenols or terpenoids. If N is highly available, then the tree will promote the vegetative growth (Brandt and Mølgaard, 2001), resulting in a decrease in the production of plant secondary metabolites. This theory is supported by experimental results showing an increase of synthesis of ascorbic acid (Brandt and Mølgaard, 2001) as a response of stress conditions such as low N availability. On the other hand, increasing nutrient availability promoted a decrease of the concentration of phenolic antioxidants (Mitchell and Chassy, 2007) and a contemporary increase of proteins (Brandt and Mølgaard, 2001).

Although the responses depend on environmental conditions, agronomic techniques, plant material, chemical nature of compounds, etc., however, from this knowledge it appears that if nutrient availability is maintained in the optimal range for the crop, no substantial difference in fruit composition should be expected in organic compared to integrated or conventional nutrient system.

Organic farming shows distinct improvement in physico-chemical and biological properties of soils determining soil quality. Soil under organic farms have lower bulk density, higher available water holding capacity, higher microbial biomass carbon and nitrogen and higher soil respiration activities compared to conventional farms managed with chemical inputs. Higher carbon and nitrogen mineralisation rates and soluble carbon content in organically managed soils indicate that sufficiently higher amounts of available nutrients are made available to the crop due enhanced microbial activity. Analysis of various organic and microbiological attributes from the soil samples drawn from organic and inorganic fertilizer using farms reveals that soil fertility status of organic farms was superior to the soils fertilized with chemical fertilizer as shown in table (6.1) below.

Table 6.1: Chemical and microbiological analysis of soil samples collected from the farmers' fields under organic and conventional systems.

Characteristics	Organic sources*Integrated nutrient**Chemical use**fertilizers***					
	Depth (cm)		Depth (cm)		Depth (cm)	
	0-7.5	7.5-15.0	0- 7.5	7.5-15.0	0-7.5	7.5-15.0
I Chemical Analysis						
pH (1:2.5)	7.25	7.25	7.41	7.43	7.51	7.51
Organic carbon (%)	0.60	0.58	0.53	0.52	0.41	0.39
Available N (kg ha^{-1})	256	255	224	222	185	184
Available P_2O_5 (kg ha^{-1})	49	49	42	41	29	28
Available K_2O (kg ha^{-1})	458	459	477	470	426	427
Mineral N (kg ha^{-1})	70.37	66.00	57.33	54.66	46.28	44.43
I Microbiological analysis						
Soil microbial biomass C (mg kg^{-1})	227	264	235	229	220	214
Soil microbial biomass N (mg kg^{-1})	39	37	34	31	30	27
Dehydrogenase activity (Ug TPF g^{-1} soil 24 hr^{-1})	54	51	45	42	35	31
Acid phosphatase activity(Ug TPF g^{-1} soil 24 hr^{-1})	629	613	603	590	558	543
Azotobacter (10^3 g^{-1})	12.7	10.5	6.3	5.3	0.9	0.6
P Solubilising bacteria(10^5 g^{-1})	9.1	8.8	6.5	6.2	3.2	2.9
Actinomycetes (10^5 g^{-1})	26.7	22.9	18.3	16.0	1.8	1.2
Flourescent pseudomonas (10^5 g^{-1})	22.3	19.9	19.9	12.1	9.9	9.1

Average of 8 soil samples, **Average of 6 soil samples, ***Average of 7 soil samples (Anonymous 2002)

Role of Soil Organic Matter

Soil organic matter, a most precious component of soil, is also considered as store house of many nutrients. It consists of a mixture of plant and animal residues in various stages of decomposition, substances synthesized chemically and biologically from the breakdown products, and microorganisms and small animals and their decomposing remains. In simple terms, it can be classified into non-humic and humic substances. Non-humic substances include those with still recognizable physical and chemical characteristics such as carbohydrates, proteins, peptides, amino acids, fats, waxes, alkanes, and low molecular weight organic acids. Most of these compounds are attacked relatively readily by microorganisms in the soil and have a short survival period. The humic substances which form the major portion of organic matter in soil are characterized by amorphous, dark coloured, hydrophilic, acidic, partly aromatic, chemically complex organic substances with molecular weight varying from few hundreds to several thousands. Humic substances are categorized into three parts:

i. humic acid which is soluble in dilute alkali but is precipitated by acidification of the alkaline extract,

ii. fulvic acid which is the humic fraction that remains in solution when the alkaline extract is acidified and

iii. humin, which is the humic fraction that cannot be extracted from the soil or sediment by dilute base and acid (Schnitzer, 1982).

Table 6.2 Physico-chemical properties of rhizosphere soil influenced by organic and inorganic fertilizers

Treatment	pH	MC%	SOC%	TN%	µ/g AP	mg/gK	mg/gSR	µ/g MBC
PC	(5.2-6.9)	(21.8-28.1)	(0.43-2.16)	(0.16-0.46)	(1.04-3.12)	(0.03-0.05)	(48.8-73.0)	(130.5-2610.6)
	5.6±0.01	24.9±0.54	1.80±0.11	0.32±0.02	1.18±0.15	0.04±0.001	65.1±2.18	1015±207.3
VC	(5.1-5.8)	(19.4-29)	(1.4-1.65)	(0.24-0.47)	(2.32-3.2)	(0.03-0.08)	(56.98-71.94)	(984.9-3803.5)
	5.4±0.05	24.24±0.9	1.50±0.02	0.31±0.01	2.66±0.07	0.05±0.005	66.11±1.54	2145.7±248.9
IPC	(5.3-6.2)	(21.4-33.9)	(1.02-2.13)	(0.20-0.53)	(1.57-2.52)	(0.03-0.06)	(52.14-72.38)	(131.5-3308.9)
	5.6±0.05	24.68±0.8	1.75±0.07	0.35±0.02	2.01±0.08	0.04±0.002	64.56±1.95	1385.1±264.4
FYM	(5.0-5.7)	(20.1-27.4)	(1.12-2.04)	(0.25-0.37)	(1.21-4.40)	(0.06-0.10)	(52.14-72.6)	(14.1-2453.09)
	4.6±0.08	23.82±0.6	1.27±0.03	0.31±0.01	2.24±0.25	0.08±0.003	56.5±1.93	940.9±240.3
CON	(4.1-5.1)	(20.4-27.4)	(1.05-1.46)	(0.19-0.44)	(1.20-3.16)	(0.041-0.063)	(34-34-71.5)	(173.0-1210.5)
	4.9±0.06	23.39±0.44	1.60±0.03	0.28±0.01	2.01±0.15	0.05±0.001	56.56±2.39	656.5±83.08
NPK	(4.5-5.3)	(20.7-26.8)	(1.43-1.86)	(0.22-0.46)	(1.84-3.72)	(0.035-0.046)	(48.84-72.38)	(86.7-1756.4)
	4.9±0.06	23.39±0.4	1.60±0.03	0.35±0.01	2.68±0.16	0.04±0.001	62.89±2.23	798.9±167.7

Note: Min and max range followed by mean ± Standard error. PC= Plant compost, VC= Vermicompost, IPC= Integrated plant compost, FYM= Farm yard manure, CON= Control, MC= moisture content, SR= Soil respiration, SOC= Soil organic carbon, TN= Total nitrogen, AP= Available phosphorous, K= Potassium, MBC= Microbial biomass carbon.
Source:Dasand Dkhar (2011).

When plant and animal remains are recycled in soil, they undergo the various stages of microbial decomposition and humification. Since agricultural soils contain little litter and decomposed litter layers, SOM generally refers to non-humic substances which constitute 10-15% of total organic materials, and the humic substances which comprise the largest fraction (85-90%). Organic matter (OM) is what makes the soil a living, dynamic system that supports all life. The significance of soil organic matter (SOM) accrues from the following facts:

➤ Organic matter is considered as a food / energy source for soil microorganisms and soil fauna. Without OM, the soil would be almost sterile and consequently, extremely infertile.

> It is the storehouse of many plant nutrients such as N, P, S and micronutrients and contributes significantly to the supply of these nutrients to higher plants. There is very little inorganic nitrogen in soils and much of it is obtained by transformation of the organic forms. Plants are therefore, dependent either directly or indirectly, for their nutritional requirement of nitrogen on SOM.

> SOM also plays an important role in improving the majority of soil physical properties such as soil structure, water 62 Winter School – CRIDA holding capacity, porosity, infiltration, soil drainage, etc.

> Soil organic matter also helps in improving various chemical properties of soil. For example, the increased cation exchange capacity and enhanced ligancy help in trapping nutrient cations like potassium, calcium, magnesium, zinc, copper, iron, etc. Improved soil buffering is it's another important contribution.

> Apart from the nutrients within the soil organics themselves, SOM contributes to nutrient release from soil minerals by weathering reactions, and thus helps in nutrient availability in soils.

> Plant growth and development are benefited by the physiological actions of some organic materials that are directly taken up by plants.

> The organic substances also influence various soil processes leading to soil formation.

Organic Matter as Soil Structure Builder and Storehouse of Nutrients:

It has been established that the organic matter content of agricultural soils is significantly correlated with their potential productivity, tilth and fertility. Although the amount of soil organic matter (SOM) in most semiarid dryland soils is relatively low ranging from 0.5 to 3% and typically less than 1%, its influence on soil properties is of major significance. Organic matter is the predominant material facilitating soil aggregation and structural stability even at low concentrations. Better soil structures helps in improved air and water relationships for root growth and in addition protect soils form wind and water erosion. The dark colour imparted by humic fraction of SOM increases the soils capacity to absorb heat and to warm rapidly in the spring. In semiarid regions with low or intermittent rainfall, organic matter is the major pool for some of the essential plant nutrients. The N, P, S contents of these soils average 0.12%, 0.05% and 0.03% respectively, with 95% of the N, 40% of the P and 90% of the S being associated with the organic matter component. Since the soil organic matter constitutes the predominant pool of plant nutrients, the decomposition and fluctuation within this pool are of major significance to nutrient storage and cycling. In many dryland cropping systems, depending on fertilizer additions and crop rotations, 50% or more of the nitrogen required by the crop comes from the mineralization of SOM. The microbial action that mediates this decomposition and nutrient release process is regulated by perturbations of the system such as wetting of dry soil, tillage, and addition and placement of residue.

These types of perturbations affect the dynamics of SOM decomposition, the size of the microbial biomass pool and nutrient release.

Role of Organic Nutrient Sources in Enhancing Nutrient-Use Efficiency with Some Typical Examples

Organic materials have a major role to play in maintaining buffering capacity of soil and are important for maintaining soil physical and biological properties. A range of factors such as soil temperature, moisture and chemical soils composition of the organic material influence N release from organic sources in soil. Nitrogen losses also occur from their use. Controlling N release from organic sources depends on their nutrient content and quality, soil properties,and the environmental and management factors(Singh *et al.*, 2001). The build-up of soil organic matter is required to increase the potential for N mineralization. The challenge in optimising crop N uptake in organic and cover crop-based systems does not entirely rely on developing organic matter pools but is more important to influence the rate and timing of N mineralisation. It was found that interactions among inputs (manure, cover crops and fertilizer) and soil organic matter influence the rate of soil N mineralisation (Horwath *et al.*, 2006). It is also well known that N from many organic fertilizers often shows little effect on crop growth in the year of application because of the slow release characteristics of organically bound N. Nitrogen immobilisation after application can occur, leading to enrichment of the soil N pool. This process increases the long-term efficiency of organic fertilizers. However, if short-term N release from organic sources is measured as mineral fertilizer equivalents, it varies from (some material) to 100% (urine). Addition of green manure along with NPK was promising during the first three seasons, while FYM along with NPK was better during later three seasons, indicating long-term use of FYM forbetter yields in rice-rice cropping systems in Vertisols (Subbaiah *et al.*, 2006). Soil fertility parameters showed conspicuous improvement over the initial status under FYM and poultry manure. Integrated nutrient management practice of using green leaf manure (*Melia*, neem and subabul) with chemical N fertilizer significantly enhanced N uptake, agronomic efficiency,apparent N recovery and yield of sunflower(Panneer and Bheemaiah. 2005). In groundnut -maize cropping system, incorporation of groundnut stover improved the subsequent maize yield than surface application. A higher N recycling efficiency was observed when residues were incorporated (Sakonnakhon *et al.*,2005). The residual effect of residue incorporation significantly improved the performance of chickpea, with improved physical and fertility parameters. Asynchrony between N released by organic materials and N demand by the crop leads to low NUE. Nitrogen use efficiency for maize could be improved when *Glyricidia sepium* prunings were incorporated 4 weeks ahead of maize planting. Addition of small doses of inorganicN fertilizer increased N uptake and yield (Makumba *et al.*, 2006). Application of fertilized green manure such as *Sesbania rostrata* proved superior as compared to paddy straw and FYM on rice yield and residual effect on linseed. Rice yield and N uptake were significantly higher under combined application of green manure and urea Non an equivalent basis. A saving of urea N up to 50%could be achieved by applying 1 t/ha (dry weight) *Sesbania rostrata* in rice (Singh *et al.*, 1999).

Chapter 7

Certification and Legislation of Organic Food

Certification is an important prerequisite for the acceptability of organic products or foods as organic by Government Regulatory Authorities, exporters, importers, as well as consumers across the world. To satisfy their requirement, a sound system of certification and labelling of the produce by a competent agency is highly essential.Certification means having the farm and the farmer's methods inspected by an accredited agency to ensure that they comply with the guidelines on the organic farming. Each certifying agency has a code of standard to prevent the marketing of substandard produce. Different certifiers use different inspection methods and criteria but the results are similar. Some certifiers use three levels of organic certification viz. Level A, the top level, are fully organic level; Level B, or the in-conversion level, is the transitional level, for produce from farms, which are being converted to organic farms. Farms in this level must meet the level A standard but are not considered organic until they have been farmed in this way for sometimes -usually at least 2 years. And Level C, or the pre-conversion level, is the period prior to the in-conversion level. The farm and its operations are under an organic inspection system, usually for a period of 12 months.

The organic certification is a procedure by which a third party between the producer and consumer gives written assurance that the product, process or service confirms to specific requirements. The farming unit for organic production has to be supervised and inspected at frequent intervals and at different stages of production before certification in order to ensure quality and authenticity. The Certification Agency has to adopt very reliable methods such as Soil tests, Water tests, Food quality tests, and other natural quantitative indicators so as to ensure credibility of the system in order to prevent fraudulent labelling of the products. It is necessary to keep the records of all management practices and materials used in organic production for five years. The crops must be grown on the land, which has been free of prohibited substances for three years prior to harvest. Crops grown on land in transition to organic (during the last three years after switching from conventional farming) cannot be labelled as *ORGANIC*. Once the produce is certified as *ORGANIC*, the producer or the processors are entitled the symbol.

Worldwide, inspection and certification of organic foods is carried out on the basis of two largely overlapping sets of guidelines and norms namely, Statutory Certification Norms and the Voluntary/Civil Certification Norms. Generally the Voluntary/Civil Certification Norms are stricter than Statutory Certification Norms. Statutory Certification Norms are legal guidelines set by Government, which is related to certification of organic produce, regulatory governing import 8 of organic foods, rules regarding equivalence between countries etc. On the other hand, various National and International forums and association such as Soil Association of UK, Organic Growers Association in various countries etc, set Voluntary/Civil Certification Norms.

The most highly accepted voluntary certifications are from agencies like CODEX, IFOAM, Naturland, Demeter, Soil Association etc. In India, Statutory Certification Norms relating to organic foods regulates the organic exports only not the domestic organic food industries. Although in India, the External certification bodies have been introduced for inspection and certification programmes since 1987. But in March 2000, the Ministry of Commerce launched the National Programme for Organic Production (NPOP), designed to establish national standards for organic products, which could then be sold under the logo *'India Organic'*. To ensure the implementation of NPOP, the National Accreditation Policy, and Programme (NAPP) was formulated, with accreditation regulations announced in May 2001 (Anonymous,2001).

These make it mandatory that all certification bodies, whether already engaged or proposing to engage in inspection and certification of organic crops and products, should be accredited by an accreditation agency. Foreign certification bodies operating in the country must also be accredited under the NAPP. For Organic Certification Agency, International Federation of Organic Agriculture Movements (IFOAM), Germany has established the IFOAM Accreditations Programme. In India, IOAM (Indian Organic Agriculture Movement), a member of IFOAM, adopted the IFOAM International Standards, the basic production standards applicable under Indian condition were prepared, and farmers growing crops as per IOAM Standards are eligible to get the Certificate and the organic label. The farmers can sale the organic produce in the local as well as International markets on the basis of IOAM label.The National Standard Committee has drafted both the concept and principles of basic standards of Organic Agriculture in 1996 in order to improve the socio economic condition of the farmers and also boost the International Trade.

At present in India, the following six authorized accreditation agencies has been approved by the Ministry of Commerce, Government of India. These include:

➤ APEDA (Agricultural & Processed Food Product Export Development Authority).

➤ Coffee Board

➢ Spices Board

➢ Tea Board

➢ Coconut Development Board

➢ Cocoa & Cashew nut Board

In addition there are four Certification Agencies accredited by APEDA such as

➢ IMO Control Pvt. Ltd., Bangalore (Institute fur Market ecologzies, Switzerland)

➢ Skal International (The Netherlands), India, Bangalore

➢ SGS (Societe Generale de Surveillance, Switzerland) India Pvt. Ltd., Gurgaon

➢ ESCOCERT (Ecological Certification, France) International, Germany

APEDA (Agricultural & Processed Food Product Export Development Authority) is an export promotion organization, involved in publicizing Indian Organic logo globally. It has also engaged to identify exclusive Agri. Export Zone (AEZ) for organic produce in some parts of country, such as organic pineapple in Tripura, where use of chemical fertilizers and pesticides is negligible.

Purpose of certification

Organic certification addresses a growing worldwide demand for organic food. It is intended to assure quality and prevent fraud. For organic producers, certification identifies suppliers of products approved for use in certified operations. For consumers, "certified organic" serves as a product assurance, similar to "low fat", "100% whole wheat", or "no artificial preservatives". Certification is essentially aimed at regulating and facilitating the sale of organic products to consumers. Individual certification bodies have their own service marks, which can act as branding to consumers. Most certification bodies operate organic standards that meet the National government's minimum requirements.

The certification process

In order to certify a farm, the farmer is typically required to engage in a number of new activities, in addition to normal farming operations:

Study the organic standards, which cover in specific detail what is and is not allowed for every aspect of farming, including storage, transport and sale.

Compliance - farm facilities and production methods must comply with the standards, which may involve modifying facilities, sourcing and changing suppliers, etc.

Documentation - extensive paperwork is required, detailed farm history and current set-up, and usually including results of soil and water tests.

Planning - a written annual production plan must be submitted, detailing everything from seed to sale: seed sources, field and crop locations, fertilization and pest control activities, harvest methods, storage locations, etc.

Inspection - annual on-farm inspections are required, with a physical tour, examination of records, and an oral interview.

Fee – A fee is to be paid by the grower to the certification body for annual survellence and for facilitatining a mark which is acceptable in the market as symbol of quality. National Project on Organic farming Deptt of Agriculture and Cooperation, Govt. of India National Centre of Organic 50 Farming, Ghaziabad.

Record-keeping - written, day-to-day farming and marketing records, covering all activities, must be available for inspection at any time. In addition, short-notice or surprise inspections can be made, and specific tests (e.g. soil, water, plant tissue analysis) may be requested. For first-time farm certification, the soil must meet basic requirements of being free from use of prohibited substances (synthetic chemicals, etc) for a number of years. A conventional farm must adhere to organic standards for this period, often, three years. This is known as being in *transition*. Transitional crops are not considered fully organic. A farm already growing without chemicals may be certified without this delay.

Table 7.1 shows some agencies involved in domestic marketing of organic produce in India

S. No	Name of the organization	Address
1.	NAVDANAYA Trust	A-60, Hauz Khas, New Delhi-110016.
2.	Devine Agro Industries Ltd.	C-9, Anoop Nagar, Uttam Nagar, New Delhi-110059
3.	Devbhoomi	Rajput Road, Dehradun, Uttaranchal
4.	Back to Nature	Near Kanak Cinema, Dehradun, Uttaranchal
5.	Mahrishi Ved Vigyan Vidyapeeth	Dunda (Kunshi), Uttar Kashi
6.	Institute of Rural Development (IIRD)	54A, Kanchan Nagar, Nakshetrawadi, Aurangabad 413002
7.	ISCON Temple	Bangalore
8.	FAB India Overseas Pvt. Ltd.	B-26, Okhla Industries Area, Phase I, New Delhi
9.	ECO-Nut Health Food Shop	J's Heritage Complex, Opp. Milk Union, P.T. Road, Kodai Kanal-624101
10.	Sresta By-products Pvt. Ltd.	203, Pavani Annexes, Road No. 2, Banjara Hills, Hyderabad 500 034, AP
11.	IOCCA	951C, 15th Cross, 8th Main, Ideal Home Township, Raja Rajeswari Nagar, Bangalore-560 098
12.	D.R. Agro Organic AS	01,, Sai Nagar, Ratnagiri, Kapadganj-387620, Gujarat
13.	Sunstar Overseas Ltd.	40 K.M. Stone, G.T. Karnal Road, Bahalgarh, Sonepat, Haryana
14.	IITC Organic India Ltd.	A-306, Indira Nagar, Lucknow-227 105

Source: Bhattacharya and Chakraborty (2005)

NSOP (National Standards for Organic Production): It has been formulated by Dept. of Commerce, Govt. of India for National Programme for Organic Production (NPOP). Any production certified as per NSOP may use the term, *"Organic"*. A product can be labelled as, *"For Export only"* when it has been produced in India to an Organic Standard other than NSOP for example EU Regulations, IFOAM etc. Truthful label claims are allowed for domestically produced organic products that meet the NSOP and an International Organic Standards. Organic Certificates remained valid for one year/until the next decision is made. The frequency of inspection is generally done once in a year. Additional inspections are conducted wherever found necessary. NSOP also formulated rules for misuse of the term, *"Organic"*. Any operation that knowingly sells per labels a product as, *"Organic"* except in accordance with the National Standards may be subject to a civil penalty India's first ever local Organic Certification Body, INDOCERT (Indian Organic Certification Agency), was established in March, 2002 with an objective to offer a reliable and affordable organic inspection and certification services to farmers, processors, input suppliers and traders. It provides certifications both for domestic as well as export market. It has been set up by a group of Indian NGO's and corporate organizations with the technical collaboration of FiBL, bio.inspecta, and the Swiss State Secretariat of Economic Affairs (SECO). INDOCERT has strong technical collaborations with two well reputed organizations from Switzerland: FiBL (Research Institute of Organic Agriculture) and bio.inspecta (the leading Swiss certification agency). Bio.inspecta assists INDOCERT for certification according to USDA national organic program (NOP) and JAS (Japanese Agricultural Standard for Organic Agriculture) through a re-certification procedure. According to the year of production, INDOCERT label the products as organic as follows,

Year wise Label

Crops	1st year	2nd year	3rd year	4th year
Annual	No label	In Conversion	Certified Organic	Certified organic
Perennials	No label	In Conversion	In Conversion	Certified Organic

Conversion to Organic Production Systems

Conversion period is actually the time between the start of organic management and certification of crops and/or animal husbandry. When traditional agricultural methods fulfil the principles of the standards, no conversion period is required. When virgin lands are used for organic purpose, no conversion period is required. The whole farm, including livestock should be converted according to standards over a period of time. If a farm is not converted at once, it should be done on a field-by-field basis. The conversion plan shall cover all aspects relevant to these standards. The converted land and animals shall not get switched back and forth between organic and conventional management. Following table shows area under conversion process.

Conversion Requirements

It should include, History and existing situation of crops, fertilizing, pest management, animal husbandry etc, a schedule for the progression of conversion and the details of the aspects which is required for change during the conversion period .Usually the conversion period is calculated on the basis of date of application to the Certificate bodies or from the date of last application of unapproved farm inputs.

Table7.2 Total areas under organic conversion process in India

S.No.	State	Total Area in ha		Total
		Organic	In conversion	
1.	Delhi	77.3	190.4	267.7
2.	Goa	5947.1	1443.67	7390.77
3.	Haryana	3585.16	5787.59	8972.75
4.	Himachal Pradesh	437.09	139.01	576.1
5.	J&K	430.63	182.44	613.07
6.	Karnataka	16099.06	35369.398	51468.458
7.	Maharashtra	105172.62	45295.12	1,50467.74
8.	Mizoram	18002.27	9857.55	27859.82
9.	Punjab	379.84	4883.77	5263.61
10.	Uttarakhand	16158.86	14906.75	31065.61

(Source: NPOP/IFOAM, 2010)

Plant products from annual production can be certified organic when the standard requirements have been met for a maximum period of 12 months before the start of production cycle. Perennial plants can be certified organic at the first harvest after at least 18 months of management according to the standards. The Certification Agencies may allow plant products to be sold as *"produce of organic agriculture in process of conversion"* when the standards requirements have been met for at least 12 months. On farms with simultaneous organic and conventional production the use of genetically engineered organisms is not permitted on the conventional part. Certification of processing units can be done when there is clear documented evidence that organic and conventional streams of processing are separated.

Chapter 8

Market Opportunities of Organic Farming

Organically produced agricultural products have received global attention in the last four years especially due to their being a multi-billion trade.World market of organic agricultural produces has been expanding continuously and estimated 25 billion US dollar in Europe and 20 billion US dollar in North America. The total market of organic food was noticed 46.2 billion US dollar of the entire world in 2007 (Anonymous, 2013). Market of organic food has been growing by 10% annually and estimated the global market price nearly 2.7 billion euro which was 1.7% (in value) of the entire food market in 2008. Different organic foods product have share like fruits and vegetables 17%, dairy products 16%, bread and flour 13% and processed food products 3% among all organic food products. In 2009, the organic agriculture had shown a sharp rise by converting 36000 new producers which were 23% more in the comparison of 2008 (Anonymous, 2013).

Table 8.1 Organic food: Indian scenario and market trends

Total Organic Area	4.43 Million Hectare
Total certified production	1.71 Million tones
Total Export	69837 MT
Value of export INR	700 Crores.

*INR=Indian Rupees. MT= Metric tons (Aivalli, 2013)

The global sales of organic food and drinks have increased by 43% from 23 billion US $ in 2002 with sales reaching 33 billion US$ in 2005. Although, organic agriculture is now present in most parts of the globe, demand remains concentrated in Europe and North America. These two regions are experiencing under supply because the production is not meeting the demand. Thus large volumes of imports are coming in from other regions. Production in developing world is rising at much faster rate than that in the industrial countries. For example, the amount of organic farm land increased in triple digits in Asia, Africa and Latin America since 2000, whereas only double digit growth has been observed in other regions. Demand for organic products mainly comes from affluent countries. Six of the G-7 countries comprise 84% of global revenues. This disparity between production and consumption of organic foods puts the industry in a fragile condition. A dip in demand from Europe and / or North America would have a major impact on global production of organic food. The industry could lose confidence as export markets close, causing oversupply and organic food prices to drop.

Organic farming has been one of the fastest-growing sectors inagriculture, and double-digit growth in sales of organic foods has provided market incentives for the U.S. agricultural sector across a broad range of products. The retail value of the organic industry grew to $31.4 billion a year in 2011, up from $21.1 billion in 2008 and $3.6 billion in 1997 (USDA, 2012). Between 2002 and 2008, acres under organic production grew by an average of 16.5 percent a year. Organic sales currently account for more than 3 percent of total U.S. food sales, and provide a larger share in categories such as produce and dairy. Growth has been particularly evident in the organic dairy sector, which accounted for 16 percent of organic sales in 2008. The number of organic milk cows on U.S. farms increased by annual average of 26 percent between 2000 and 2008. As demand for organic food has increased, the U.S. agricultural sector has taken steps to meet it; the number of operations certified as organic grew by 1,109—or more than 6 percent—between 2009 and 2011. The USDA has taken steps both to promote and to regulate the growing organic food industry by establishing the National Organic Program (NOP), whose mission is to ensure the integrity of USDA-certified organic products in the United States and throughout the world. The NOP accredits nearly 50 domestic organic certifying agents who are authorized to issue an organic certificate to operations that comply with the USDA organic regulations. Between 2009 and 2011, the USDA has also supported its own

Scientists and university researchers with more than $117 million in funding focused on improving the productivity and success of organic agriculture. For example, USDA research on weed management for organic vegetable production has produced techniques and tools that can help control 70 percentof weeds at 15 percent of the previous cost for weed control. The increasing demand for organic foods has been accompanied by a growing "local" movement. The markets for organic and local food regularly overlap: organic farmers are much more likely than conventional farmers to sell their products locally with about a quarter of all organic sales in 2004 made within an hour's drive of the farm (Greene *et al.* 2009). Similarly, 82 percent of all farmers' markets had at least one organic vendor. Sales of locally produced foods make up a small but growing part of U.S. agricultural sales, particularly for small farms. The USDA estimates that the farm-level value of local food sales totaled nearly $5 billion in 2008, or 1.6 percent of the U.S. market for agricultural products. An estimated 107,000 farms, or 5 percent of all U.S. farms, are engaged in local food systems, with small farms (those with less than $50,000 in gross annual sales) accounting for 81 percent of all farms reporting local food sales in 2008 (Low and Vogel, 2011). Examples of the types of farming businesses that are engaged in local foods are direct-to-consumer marketing, farmers' markets, farm-to-school programs, community-supported agriculture, community gardens, school gardens, food hubs and market aggregators, kitchen incubators, and mobile slaughter units, among a myriad of other types of operations.

The Government of India has also launched the National Programme for Organic Production (NPOP) in the year 2001. The NPOP standards for production and accreditation system have been recognized by the European Commission and Switzerland as equivalent to their country standards. Similarly, the United States Department of Agriculture (USDA) has recognized NPOP conformity assessment procedures of accreditation as equivalent to those in the US. With these recognitions, the Indian organic products duly certified by the accredited certification bodies of India are accepted by the importing countries (Reddy, B. S. 2010).India produced around 5, 85,970 Mt (Table 8.2) of certified organic products including all varieties of food products.India exported 86 items in the year in 2007- 08 —the total volume being 37533 Mt.

Table 8.2: Status of organic food production in India

Total area under certified organic	2.8 M ha
Total production	585970 M tonnes
Total quantity exported	194560 M tonnes
Value of total export	Rs. 30, 124
Number of farmers	141904

About 50% of the organic food production in India is targeted towards exports, there are many who look towards organic food for domestic consumption. Organic food is priced over 25% more than conventional food in India. But now since organic food has been declared as completely safe for domestic consumption, many parents are willing to pay this higher premium due to the perceived health benefits of organic food. The increase in organic food consumption in India is evident from the fact that many organic food stores are mushrooming in India. Today organic food is an essential part of many retail food stores and restaurants. The pattern of organic food consumption in India is much different than in the developed countries. However, the Indian organic food consumer needs education. There are many consumers who are unaware of the difference between natural and organic food. Many people purchase products labeled as Natural thinking that they are Organic. As far as consumption of organic food export is concerned, Organic food exports from India are increasing with more farmers shifting to organic farming. India has now become a leading supplier of organic herbs, organic spices, organic basmati rice, etc.

India exported 86 items in the year in 2007- 08 —the total volume being 37533 Mt. The export realization was around US $ 100.4 million, registering a 30 per cent growth over the previous year. Organic products are mainly exported to EU, US, Australia, Canada, Japan, Switzerland, South Africa and the Middle East countries. Cotton leads among the products exported (16, 503 Mt) (Reddy, B. S. 2010). The states of Uttarakhand and Sikkim have declared their states as - organic states'. In Maharashtra, since 2003, about 5 lakh ha area has been under organic farming (of the 1.8 crore ha of cultivable land in the state). In Gujarat, organic production of chickoo, banana and coconut is being encouraged both from profit as well as yield point of view. In Karnataka, the area under non certified organic farming (4750 hectares) was substantially high as comparison to land was under certified organic farming (1513 hectares). The reasons behind this transition of shifting towards organic farming are sustained soil fertility, reduced cost of cultivation, higher quality of produce, sustained yields, easy availability of farm inputs and reduced attacks of pest and diseases. Apart from this, the government of Karnataka had released a state organic farming policy in 2004 for encouraging organic farming. Infact, most of the northeastern states are also encouraging organic farming. In Nagaland, 3000 ha area is under organic farming. Also States like Rajasthan, Tamil Nadu, Kerala, Madhya Pradesh, Himachal Pradesh and Gujarat are promoting organic farming vigorously. Various farmers 'organizations have been established in different states for the marketing of organic products. For example, the establishment of the - Chetana' in three states: Andhra Pradesh (Asifabad and Karimnagar), Maharashtra (Vidarbha, Akola and Yavatmal) and Tamil Nadu (Dindigul and Tuticorn). However, there are indeed some constraints being faced by the farmers for transforming their conventional farming system into organic farming system. Lanting L. (2007) has identified some of the problems as follows: Non-payment of premium price for these products because they are in the transition stage, lack of storage facility, with cash paid (preferably 70% of the crop value) for the stored products. Here the urgency for the assistance from the government as a helping hand is of utmost importance for overcoming the barriers faced due to the transition from conventional farming to organic farming. The production of different commodity under organic management is given in table 8.3 below.

Table 8.3: shows production of commodities under Organic Farming

S.No	Commodities	Quantity produced in MT		Total
		Organic	in conversion	
1.	Rice	44335	32354	76690
2.	Wheet	6892	15364	22560
3.	Pulses	17560	16785	34345
4.	Fruit and Vetetables	194505	538073	7,32579
5.	Herbal/medicinal plant	129543	58767	1,88310

Table 8.4: Sales and export of organic food in India

Product	Domestic sales,	MT Export, MT
Rice	5000	6877
Wheat & Flour	3000	-
Pulses	2500	-
Tea	1500	2928
Coffee	750	320
Spices	500	-
Fruits and Vegetables	5000	1639*

MT=Metric tons   Source: Aivalli, (2013).

Conditions for Products used in fertilization and soil conditioning in organic farming

Items	Conditions for use
Material from plant and animal origin	
Matter produced on an organic farm unit	
Farmyard and poultry manure, slurry, urine	Permitted
Crop residues and green manure	Permitted
Straw and other mulches	Permitted
Composts and Vermicompost	Permitted
Matter produced outside the organic farm unit	
Blood meal, meat meal, bone meal and feather meal without preservatives	Restricted
Compost made from plant residues and animal excrement	Restricted
Farmyard manure, slurry, urine	Restricted
Fish and fish products without preservatives	Restricted
Guano	Restricted
Human excrement	Restricted
Wood, bark, sawdust, wood shavings, wood ash, wood Charcoal	Restricted
Straw, animal charcoal, compost and spent mushroom	
and vermiculate substances	Restricted
Compost from organic household	Restricted
Compost from plant residues	Restricted
Sea weed and sea weed products	Restricted
By products from the industries	
By-products from the food and textile industries of biodegradable material of microbial, plant or animal origin without any synthetic additives	Restricted
By products from oil palm, coconut and cocoa (including fruit bunch, palm oil mill effluent, cocoa peat and empty cocoa pods.	Restricted
By-products of industries processing ingredients from organic agriculture	Restricted
Extracts from mushroom, Chlorella, Fermented	
product from *Aspergillus*, natural acids (vinegar)	Restricted
Mineral Origin	
Basic slag	Restricted
Calcareous and magnesium rock	Restricted
Lime, limestone, gypsum	Permitted
Calcified sea weed	Permitted
Calcium chloride	Permitted
Mineral potassium with low chlorine content (e.g. sulphate of potash, kainite, sylvinite, patenkali)	Restricted

Items	Conditions for use
Natural phosphates (rock phosphate) Restricted Trace elements	Permitted
Sculpture	Permitted
Clay (bentonite, perlite, zeolite)	Permitted
Microbiological origin	
Bacterial preparations (biofertilizers)	Permitted
Biodynamic preparations	Permitted
Plant preparations and botanical extracts	Permitted

Conditions for Products used in Plant pest and disease control

Items	Conditions for use

Material from plant and animal origin Plant based repellents

(Neem preparations from *Azadirachta indica*)	Permitted
Algal preparations (gelatin)	Permitted
Casein	Permitted
Extracts from mushroom, chlorella, fermented products from	Permitted
Aspergillus Propolis	Restricted
Beeswax, Natural acids (vinegar), plant oils, Quassia	Permitted
Rotenone from *Derris elliptica*, *Lonchocarpus*, *Tephrosia* spp.	Restricted
Tobacco tea (pure nicotine is prohibited)	Restricted
Preparation from *Ryania* species	Restricted

Mineral origin

Chlorides of lime/soda	Restricted
Burgundy mixture	Restricted
Clay (bentonite, perlite, vermiculite, zeolite)	Permitted
Copper salts/ inorganic salts (Bordeaux mix, copper	
hydroxide, copper oxychloride)	Not alllowed
Quick lime	Restricted

Mineral origin

Diatomaceous earth	Permitted
Light mineral oils	Permitted
Permangnate of potash	Restricted

Insects origin

Release of parasites, predators of insect pests	Restricted
Sterilized insects	Restricted
Sterilized insect malesnot allowed Microorganisms used for biological pest control	
Viral, fungal and bacterial preparations (bio-pesticides)	Restricted

Items	Conditions for use
Others	
Carbon dioxide and nitrogen gas	Permitted
Soft soap, soda, sulphur dioxide	Permitted
Homeopathic and ayurvedic preparations	Permitted
Herbal and biodynamic preparations	Permitted
Sea salt and salty water	Permitted
Ethyl alcohol Traps, barriers and repellants Physical methods (*e.g.* chromatic traps,	Not allowed
mechanical traps)	Permitted
Mulches, nets	Permitted
Pheromones – in traps and dispensers only	Permitted

Scope of Organic Farming in Jammu and Kashmir

Some of the key concerns and opportunities of agriculture in Jammu and Kashmir State, make a good case for promoting organic agriculture. The mountain/ hill farmlands being poor in organic carbon require ways to supplement it for sustainable agriculture. Presently farmers have been making use of chemical fertilizers to maximise production on the farmlands. Soils are already in intense use for vegetable farming and orchards are showing the fatigue factor, indicated by rising need for more inputs, lesser production and increasing incidence of crop diseases on these croplands.

Farmers in vegetable growing areas and apple growers are in a dilemma about available inferior options – and moving towards organic practices of soil fertility management is the answer.J & K mostly a hilly terrain, its majority of the area suffers from massive soil erosion and run-off losses especially in Jammu and Ladakh region. Besides soil erosion and run-off losses, more than 70% of the state's arable land is under un-irrigated conditions where cultivation of crops alone is very risky (Gupta *et al.*, 2005). As per the latest survey, area bought under organic farming in Jammu and Kashmir is 3239 hectares and about 613.07 ha have been certified. (Anonymous, 2010).The cropped area is least or no dependent on chemical fertilizers, pesticides especially in hilly areas (Kandi belt) of the state. The varied agro-climatic zones with potential of diversification and availability of considerable amount of organic crop residues and wastes also makes possible to convert atleast these areas into organic farming very easily.

The fruit farmers of Kashmir valley are today on toes to increase use of pesticides, which not only is increasing the cost of production but the quality of produce and food safety both are compromised. Organically grown fruits will have better quality; will get better prices (premium prices in case of certified organic).The shapes of fruit business of Kashmir Valley can change drastically by adopting organic ways of farming. Farmers so are benefit from premium value of niche crops, such as fruits crops like apples, cherry, pear and walnut, Mushkbhuji and Kamad fragrant rice varieties of Kashmir Valley, Kirkichoo apples and indigenous apricots of Kargil and Leh and Basmati rice and red rice varieties grown in parts of Jammu region.Theniche value of these commodities will be further enhanced by producing them organically.

A new thinking is developing, whereby the organic agriculture is being viewed as a precursor to dynamic change, for an otherwise stagnant agriculture sector. There is wider acknowledgement of the fact that in hilly areas we should put thrust on promoting organic farming for the larger benefit of farming consumers and the environment.

Limitations of Organic Farming

1)Low Amount of Plant Nutrients

The organic sources like FYM, compost, poultry manure , straw, etc contain very low amount of plant nutrients (both macro and micro), not sufficient to meet the nutrient requirements of crops. On the other hand, large amounts of nutrients are needed to fulfill the nutrient requirements of crops for higher yields.

2) Release of Nutrients

In organic farming the use of chemical fertilizers is not allowed and the nutrient demand of crop plants has to be met with organic manures which are not only low in essential plant nutrients but also release them slowly especially when temperature remains low for most of the growing period of crop as compared to other parts of the country. This makes the synchronization of nutrient release from organic manures and their uptake by crop plants a difficult task. Thus, supply of plant nutrients through organic manures requires much more skill on the part of the farmers.

3) High C: N Ratio of Different Organic Residues

Although a number of organic residues like paddy straw, wheat straw, sugarcane trash, etc can be used to make up the plant nutrients need yet they have wide C: N ratio. If they are not fully decomposed and added as such, will cause immobilization of nutrients, especially of nitrogen. The yield of crops will be affected badly.

4) Low Yield of Crops

With the use of organic sources of nutrients, the yield of the crop is very low especially during initial stages, although it becomes stabilized later on yet complete dependence on pure organic farming will not be sustainable in the long run.

5) Control of Weeds

Organic sources of nutrients promote profuse proliferation of weeds that compete with the plants for different nutrients, space, light, water etc. On the other hand weedicides cannot be used in organic farming, and thus economic weed control remains a challenge in organic farming.

6) Scarcity of Biomass

Readily available and abundant organic sources are not enough to meet the requirements of composts, vermicomposts etc. A lot of organic wastes and plant residues get lost in the fields being difficult to collect and handle or use properly. Moreover, people in rural areas use plant residues like straw, stubbles, prunned material and cow dung for producing charcoal, an as energy source or as a fuel for cooking purposes etc. because dependence on other energy sources like LPG, diesel, petroleum products etc proves costly to afford.

7) High Input Costs

Local or farm renewable organic resources like neem cakes, groundnut cakes, silt, cow dung, earthworms, etc are becoming costly day by day than the conventional or industrially produced chemical fertilizers & pesticides e.g, earthworms are sold at 1800 Rs/kg in the market. Chemical fertilizers are easier to purchase given the farmer has purchasing power.

8) Lack of Target (Institutional) Groups

The target groups of organic food produced include big hotels, restaurants, airlines, cafes, etc which can afford to pay higher prices (premium prices) for high quality organic foods but these are lacking in the state. Common people cannot afford to pay higher prices for organically produced food.

9) Small Holding & Poor Farmers

Most farmers in Jammu & Kashmir are outlined as small holdings and poor. They are not directly connected to markets to buy or sell food. Since organic farming's main attraction is export, small farmers are less able to compete when the international trade brings down prices even in local markets and in some cases even to survive. Thus, small farmers' interest can be safeguarded in organic farming is an issue to be carefully tackled.

10) Market &Infrastructural Problems

There is tremendous lack of infrastructural facilities for processing, packing, storage etc. to avoid contamination of organically produced food and to meet the organic standards. Daily consuming commodities like vegetables, fruits, milk, etc. must be continuously supplied to the markets as their demand is high. However, due to lack of streamlined marketing and distribution network, such costly organic foods may stale in the way before being consumed.

11) Lack of Awareness

There is lack of awareness and knowledge about modern methods or techniques of composting, vermicomposting etc from both making as well as application point of view among the farmers and thus both quality and efficacy are poor at the end. Also the guidelines for production, processing, transportation and certification are beyond the understanding of a common farmer.

12) Certifying Oriented Problems

Also before producing marketable products an organic farm has to have a transition period of 1 to 3 years depending upon the certifying agency's requirements and during this period the farmers have to grow the crops as per standards set for organic farming and thus produce about 3/4th of the normal yield. Yet they have to market the produce in the open general market. Small and marginal farmers can hardly afford to do so. Moreover the farmers won't opt for certification because of the costs involved as well as the extensive documentation that is required by certifiers.

13) Lack of credit at low interest rates

The farmers in areas with little or low consumption of agro-chemicals and fertilizers and those in rainfed areas generally have very little credit facilities in the region and being poor and small holding these farmers cannot afford or find all the necessary inputs to get high yields.

Conclusion

Keeping in view the aforesaid advantages and limitations of organic farming with respect to Jammu and Kashmir, following generalisations can be made in the context of farming.

Wherever it possible, organic farming can be encouraged especially in horticultural crops where its effects have been found better. For example in Kashmir, walnut cultivation is purely organic based horticultural crop and of high quality has tremendous export potential in the international market. Similarly, almond can be grown with no use of chemical fertilizers. So the areas under walnut cultivation can be considered as 'organic by default'.

Since Jammu and Kashmir is mostly a hilly state, there are a number of farms in the uplands especially in the Kandi belts where use of chemical fertilizers, pesticides is practically very less and farmers depend entirely on organic sources of nutrients for production. For example potato cultivation in Heripora (Shopian) and Larnoo (Ang.) and Kaala zeera production in Gurez (Bandipora). So these areas can be considered as 'relatively organic'. The farmers in these areas use very low fertilizers or traditional organic nutrient sources either due to their belief or economic reasons.

'Floating gardens' or 'Floating islands' of Kashmir formed from weeds of Dal Lake are of much importance for the production of vegetables. In these gardens, most versatile organic manure added by houseboat owners is the *hydrilla* muck. Water hyacinth and other aquatic weeds are also used. Not even a single fertilizer is added for vegetables production. Although the use of chemical fertilizers was tried but instead of increasing the yield of vegetables, it was reduced probably due to presence of more amount of available nutrients. However, vegetable production in these gardens cannot be sustainable at a large scale that too for commercial purposes, keeping environmental conservation and heritage preservation (Dal Lake) in consideration.

Similarly, the vegetable production in Leh and Kargil districts of Ladakh region is based on organic fertilizers like processed night soil compost. Use of chemical fertilizers is hardly preffered for vegetable production in this region because of greater irrigation needs.

In Jammu and Kashmir, some areas like Reasi, Arnas, Tanda, Pouni, Ghordi, and Tikri falling in (Udhampur) Jammu region are considered best for organic ginger production. Maize-ginger intercropping system with an application of FYM @ 20-50 tonnes /ha is commonly practised in some parts of Kathua district. The spices have high export potential.

Certain fruits and vegetable crops where use of higher doses of chemical fertilizers (especially N may lead to higher NO_3 content) and imbalanced nutrition of crops occurs should be grown on organic sources. Mushroom production through organic methods is achieved by substrate selectivity, steam pasteurisation and use of bio-pesticides against parasitism of fruiting bodies or mycelium. Some plantation crops where nutrient removal is very less and recycling of these through leaf fall is high, can be grown as organic.

Soils in Jammu and Kashmir have inherently good nutritional status being mostly Alfisols and slightly alkaline but low in organic matter. Soils having fixation capacity of nutrients where nutrient availability is very low even on the application of chemical fertilizers like strong alkaline, acidic or calcareous, should be sufficiently supplied with organic fertilizers.

Forests are the best examples of natural farming. No one irrigates them, no chemical fertilizers or pesticides are used or any cultural practice is done, yet forests supply readily all kinds of fuel, timber, fodder and non-wood forest products (NWFP) like gums, spices, herbs, edible nuts, aromatic plants, etc in a sustained manner. Thus forests are beautiful examples of natural farming.

Use of pesticides be minimised in case of various fruits, rather neem-based and other botanical pesticides should be used to control Sanjose scale in case of apple. Chemicals in neem like azadirachtin, nimbin, nimbidin, nimbidic acid, quercitin and thionemone are responsible for toxicity of root exudates and other plant parts. Neemacide is used against potato beetle, desert locust, grasshopper and moth. Similarly,soil amended with castor leaves was found effective against *M. javanica.*

Flowers like lily, marigold, tulip and some aromatic & medicinal plants like Lavender, Geranium, etc can be developed as such without even organic fertilizers in gardens which can generate employment for youth through entrepreneurship, besides having good export potential for their quality. Organic Lavender (*Lavendula officinalis*) best oil yielding and versatile industrial crop can be ideally cultivated in hilly areas of state. Organic Geranium for oil is also cultivated for its high export potential. However, economic weed control and disease management remains a challenge in these plants under organic production.

Bee-keeping for various products mostly for honey and wax and also for pollen, propolis, and royal jelly, a recent phenomenon in apiculture growing in response to the global changes in bee-keeping management, particularly the spread for *varroa* mite and its treatment, can be managed under organic sources.

From aquaculture food products like Trout for fish meal, oils, proteins can be grown on organic sources but nutrition, holding facilities, and post-harvest are the technical issues in organic fish farming.

Epilogue

Organic Farming has the twin objective of the system sustainable and environmental sensitivity. In order to achieve these two goals, it has developed some rules and standards which must be strictly adhered to. There is very little scope for change and flexibility. Thus, the Organic Farming does not require best use of options available rather the best use of options that have been approved. These options are usually more complex and less effective than the conventional system .With ever increasing population having huge food requirements and meagre availability of organic resources, lack of institutional and target groups, high input costs, marketing and infrastructural problems, small holding and poor farmers, certifying oriented problems etc., 'pure organic farming' is not possible; rather some specific area can be diverted to organic farming for export of high quality horticultural , plantation crops in the context of farming in Jammu and Kashmir. In this context, it will relevant to quote Nobel Laureate Dr. Norman Borlaug (2002) who said that," Switching on food production to organic would lower crop yields. We can use all the organic that are available but we are not going to feed six billion people with organic fertilizers." For agriculture, use of chemical fertilizers cannot be totally eliminated, rather can be reduced, or minimized.

It has been proved by various experiments that by conjoint application of inorganic fertilizers along with various organic sources are capable of sustaining higher crop productivity, improving soil quality and soil productivity, besides supplying N, P and K, these organic sources also helps in alleviating the increasing incidence of deficiencies of secondary and micronutrients. The commercial mineral fertilizers will have to be bearing the main burden of supplying plant nutrients to meet the nutrients to meet the food requirements of increasing populations. Therefore, these organic resources should be used in integration with chemical fertilizers to narrow down the gap between addition and removal of nutrients by crops as well as sustain the quality of soil and achieve higher crop productivity. Complete adoption of "pure Organic Farming" is not possible due to its high cost, unavailability of organic resources, productivity, social and economic reasons etc which will leave many more people hungry. In this context, renowned Agricultural scientist and thinker Dr. M.S. Swaminanthan (2003) said that," a hungry man is an angry man" and," if the hungry man happen to a young man, then we have a potential terrorist amongst us" as stated by eminent Scientist Prof. Chhonkar (2003). Thus, adoption of Pure Organic Farming is possible partially, more specifically crops having high export potential in International markets .On the other hand full adoption of Integrated Green Revolution Farming, another option of Organic Farming can be possible to a large extent, where, the basic trends of the green revolution such as intensive use of external inputs, increased irrigation, development of high yielding and hybrid varieties as well as mechanizations of labour are retained with much greater efficiency on the use of these inputs with limited damage to the environment and human health. For this purpose some organic techniques are developed and combined with the high input technology in order to create Integrated Systems such as, "Integrated Nutrient Management" (INM), "Integrated Pest Management" (IPM) and biological control methods which reduce the need for chemicals.

References

Abu Saleha and Shanmugavelu, K.G. 1988.Effect of organic vs. inorganic sources of nitrogenon growth, yield, and quality of Okra.*Ind.J.Hort.*29: 312-318

Adjei-Nsiah, S., Kuyper, T.W., Leeuwis, C., Abekoe, M.K., Cobbinah, J., Sakyi-Dawson, O. and Giller, K.E. 2008.Farmers' agronomic and social evaluation of productivity, yield and N_2-fixation in different cowpea varieties and their subsequent residual N effects on a succeeding maize crop.*Nutrient Cycling Agro-ecosystems* **80:**199-209

Aivalli G. 2013. Indian - organic food market. Yes bank limited, Jaivik India. (http://www.jaivikindia.com/presentations/session-1/Girish-Aivalli.pdf).Access date: 30 Sept, 2013.

Alam, A. and Wani, S.A. 2003 Emerging need for organic agriculture and strategies for its optimization. In souvenir of National Seminar on organic products and their future prospects, organized at SKUAST-K, Shalimar, Srinagar,Jammu& Kashmir 2003.

Anonymous 2001. Report of Task Force on Organic Farming, Department of Agriculture and Cooperation, Ministry of Agriculture, Government of India, 2001, p. 76.

Anonymous 2002.Annual Report.AICRP on Microbiological Decomposition and Recycling of farm and City Wastes, Indian Institute of Soil Science, Bhopal.

Anonymous, 2010.Report of Task Force on Organic Farming, Department of Agriculture and Cooperation, Ministry of Agriculture, Government of India.

Anonymous, 2011.Annual Report (Kharif), Directorate of Agriculture Government of Jammu and Kashmir.

Anonymous 2013. Organic Agriculture: the return to nature. Access date: 01 Oct,2013 .http://www.rungismarket.com/en/bleu/enquetesrungisactu/Agriculturebiol ogique654.asp.

Alam, A. and Wani, S.A. 2003 Emerging need for organic agriculture and strategies for its optimization. In souvenir of National Seminar on organic products and their future prospects, organized at SKUAST-K, Shalimar, Srinagar,Jammu & Kashmir, 2003.

Asami, D.K., Hong, Y.J., Barret, D.M. and Mitchell, A.E. 2003.Comparison of the totalphenolic and ascorbic acid content of freeze-dried and air-dried marionberry,strawberry, and corn grown using conventional, organic, and sustainable agriculture practices. *J. Agri. Food Chem.* 51:1237-1241.

Badgley, C., Moghtader, J., Quintero, E., Zakem, E., Chappell, M. J., Aviles-Vazquez, K.,Samulon, A., Perfecto, L. (2007): Organic agriculture and the global food supply. *Renew. Agric. Food Syst.*, 22, pp. 86–108.

Balesdent, J. and Balabane, M. 1996. Major contribution of root to soil, carbon storage inferred from maize cultivated soils. *Soil Biol. Biochem.* 28:1261-1263

Banik S. and B. K. Dey. 1982. Available phosphate content of an alluvial soil as influenced by inoculation of some isolated phosphate solubilizing microorganisms. *Plant Soil* 69:353-364.

Bhadoria, P.B.S.; Prakash, Y.S. and Rakshit, A. 2002.Importance of Organic Manures in Improving Quality of Rice and Okra.*Environment and Ecology*, 20(3): 628-633.

Bhattachary, P. and Mishra, U.C.1995.*A book on biofertilizer for extension workers published by the National Biofertilizer Development Centre, Ghaziabad.*

Bhattacharyya, P. and Verma, V. K. 2005.Current status of certifying agencies in India. In: Souvenir: National Seminar on National Policy on promoting Organic Farming. National Centre of Organic Farming, Ghaziabad.40-43.

Bhattacharya, P. and Chakraborty, G. 2005. Current status of organic farming in India and other countries.*Indian Journal of Fertilizers* 1(9):111-123.

Borlaug, N.E. 2002. CNS News com. May 01: 2002, (*http://www.scientificalliance.com/ news,organic_food/organic_forests.030502.html* .

Bockstaller, C., Girardin, P., van der Verf, H.M., 1997. Use of agro-ecological indicators for the evaluation of farming systems. *European Journal of Agronomy* 7, 261±270.

Bowler, I., 1992.Sustainable agriculture as an alternative path of farm business development. In: Bowler,Bryant, Nellis (Eds.), Contemporary Rural Systems in Transition. CAB International, New York.

Brady, N.C. 1996. The nature and properties of soils.10th Ed. Prentice Hall of India, Private Limited, New Delhi 620 p.

Brandt, K. and Mølgaard, J.P. 2001. Organic agriculture: does it enhance or reduce the nutritional value of plant foods?. *J. Sci. Food Agri.* 81: 924-931.

Brown, A.W.A. 2008.Insect control by chemicals, John Wiley & Sons, New York.

Brown, B.D., Gibson, R.C., Geary, B. and Morra, M.J. 2008.Bio fumigant biomass, nutrient content and glucosinolate response to phosphorus. *Journal of Plant Nutrition* 31: 743-757.

Carbonaro, M., Mattera, M., Vicoli, S., Bergamo, P. and Cappelloni, M. 2002.Modulation of antioxidant compounds in organic vs conventional fruits (peach *Prunuspersica* L., and pear, *Pyrus communis* L.). *J. Agri. Food Chem.* 50:5458-5462

Carter, V., Dale, T., 1974.Topsoil and Civilisation.University of Oklahoma Press, Norman, OK.

Caroll, C.R., Risch, S., 1990. An evaluation of ants as possible candidates for biological control in tropical annual agroecosystems. In: Gliessman, S.R. (Ed.), Agroecology. Researching the Ecological Basis for Sustainable Agriculture. Springer-Verlag, New York, pp. 30±46.

Chhonkar, P.K. 2003. Organic farming: Science and belief. Dr R. V. Tamhane Memorial Lecture delivered at the 68th Annual Convention the Indian Society of Soil Science, CSAU&T, Kanpur, 5[th]November 2003.

Clark, M.S., Ferris, H., Klonsky, K., Lanini, W.T., Bruggen, A.H.C., and Zalom, F.G. 1998. Agronomic, economic, and environmental comparison of pest management in conventional and alternative tomato and corn systems in northern California.*Agriculture Ecosystems and Environment***68**: 51–71.

Codex Alimentarius Commission, 2001. Guidelines for the Production, Processing, Labelling and Marketing of Organically Produced Foods. First Revision.Joint Food and Agriculture Organisation (FAO) and World Health Organisation (WHO) Food Standards Programme, Rome, Italy.Available at Websitehttp://www.codexalimentarius.net/download/standards/360/CXG_032e.pdf (verified 15 October 2009).

Collings, H.G. 1955. Commercial Fertilizers – *Their sources and use*, McGraw Hill Book Co., New York CRIDA. 1998. Annual report, Central Research Institute for Dryland Agriculture, Hyderabad p.69-70.

Das, Bibhuti. B and Dkhar,M. S. 2011. Rhizosphere Microbial Populations and Physico Chemical Properties as Affected by Organic and Inorganic Farming Practices. *American-Eurasian J. Agric. & Environ. Sci.,***10** (2): 140-150,

Dutton, V. M. and C. S. Evans. 1996. Oxalate production by fungi: its role in pathogenicity and ecology in the soil environment. *Can. J. Microbiol.* 42:881-895.

Dwivedi, V. 2005. Organic farming: Policy initiatives. In: Souvenir: National Seminar on National Policy on promoting Organic Farming. National Centre of Organic Farming, Ghaziabad. P58-61.

Escobar, M. E. O., Hue, N. V. (2007): Current development in organic farming. In: Pandalai,S. G., (Eds.), Recent Research Development in Soil Science 2, Research Signpost, Kerala, India. pp. 29-62.

Fairweather, J.R., Campbell, H., 1996. The decision making of organic and conventional agricultural producers (AERU Research Report No. 233). Lincoln University, New Zealand.

FAI, 2004. Fertilizer Statistics (2003-04). The Fertilizer Association of India, New Delhi.

FAO, 2007. Codex Alimentarius – Organically Produced Foods, FAO, Rome Fertilizer statistics 2003-04, The Fertilizer Association of India, New Delhi. pp.77.

FAOSTAT. 2009. FAO Statistical Database Domain on Fertilizers: Resource STAT Fertilizers. Food and Agriculture Organisation of the United Nations (FAO) Rome, Italy.Available at Web site http://faostat.fao.org/site/575/default.aspx#anchor (accessed 7 October 2009).

Funtilana, S. 1990. Safe, inexpensive, profitable, and sensible. *International Agricultural Development*, March-April 24.

Geier, B. 1999.International federation of organic agriculture movements, in sustainable agriculture solutions.The action report of the sustainable agriculture initiative, Novello Press, London, UK.

Giovannini, D., Scudellari, D., Aldini, A. and Marangoni, B. 2001. Esperienze diconduzione del terreno in un pescheto biologico. *Frutticoltura* 1:21-29.

Glenn, N.A., Pannell, D.J., 1998. The economics and application of sustainability indicators in agriculture. Paper presented at the 42nd Annual Conference of the Australian Agricultural and Resource Economics Society, University of New England, Armidale: January 19±21, 1998.

GoI, (Government of India). 2008b. APEDA. Ministry of Commerce and Industry, New Delhi, India.

GOI, 2001.The report of the working group on organic and biodynamic farming, Planning Commission, Government of India. pp:1-25.

Goldstein, A. H. 1995. Recent progress in understanding the molecular genetics and biochemistry of calcium phosphate solubilization by Gram-negative bacteria. *Biol. Agri. Hort.* 12:185-193.

Gomez, A.A., Kelly, D.E., Syers, J.K., Coughlan, K.J., 1996. Measuring sustainability of agricultural systems at the farm level.Methods for assessing soil quality. SSSA Special Publication 49, 401±409.

Grewal, J.S. and Jaiswal, V.P. 1990. Agronomic studies on potato under AICRP, *Tech.Bull.*20, CPRI, Shimla, pp 1-120 Gupta, R.D. Kher, Deepak and Jalali, V.K. 2005. Organic Farming: Concept and Prospective in Jammu and Kashmir. *Journal of Research, SKUAST-J* **4**: 25-37.

Hall, B., 1996. Posting to the Sanet-Mg Sustainable Agriculture Internet Discussion List, 6 February 1996.

Heffer, P. and Prud'homme, M. 2013. Short-Term Fertilizer Outlook.International Fertilizer Industry Association (IFA), Agriculture.

Henning, J., Baker, L., Thomassin, P., 1991. Economic issues in organic agriculture. *Canadian Journal ofAgricultural Economics* 39, 877±889.

Ikerd, J., 1993. Two related but distinctly di€erent concepts: organic farming and sustainable agriculture. *Small Farm Today* 10 (1), 30±31.

Hilda, R. and R. Fraga. 2000. Phosphate solubilising bacteria and their role in plant growth promotion. *Biotech. Adv.* 17:319-359.

Hinsinger, P. 2001. Bioavailability of soil inorganic P in the rhizosphere as affected by root induced chemical changes: a review. *Plant Soil* 237:173-195.

Hodge, I., 1993. Sustainability: putting principles into practice. An application to agricultural systems.Paper presented to 'Rural Economy and Society Study Group', Royal Holloway College, December 1993.

Horwath, W., Kabir, Z., Reed, K., Kaffka, S., Miyao, G. and Kent.2006. Long-term assessment of N use and loss in irrigated organic, low-input and conventional cropping systems. 18[th]World Congress of Soil Science, held during 9–15 July,Philadelphia, Pennsylvania, USA.

Howard, S. A. 1940. An agricultural testament, Research Foundation for Science, Technology and Ecology, New Delhi, India.

IFOAM, 1998.Basic Standards for Organic Production and Processing. IFOAM Tholey-Theley,Germany.

IFOAM, 2007.Organic agriculture worldwide - Directory of the member organisations and associates of IFOAM, 2006/2007.International Federation of Organic Agriculture Movements. Germany.

IFOAM, FiBL & SOEL, 2009. About 31 million certified organic hactares worldwide. Federation of Organic Agriculture Movements (IFOAM), Swiss Research Institute of Organic Agriculture (FiBL) and Foundation Ecology and Farming (SÖL), Germany. www.ifoam.org/press/press/pdfs/pm-weltweit-englisch.pdf.

IGNOU, (2007):. BAPI-003 Economics and marketing of Organic produce. http://vedyadhara.ignou.ac.in/wiki/images/b/bb/BAPI-003-01.pdf (Last accessed on September 2011).

Katan, J., 1996. Cultural practices and soilborne disease management. In: Utkhede, R.S., Gupta, V.K. (Eds.), Management of Soilborne Diseases. Kalyani Publishers, New Delhi, pp. 100–122.

Katznelson, H. and Bose, B 1959. Can. J.: Microbiol. 5, 79-85 L.(2004): XII. Int. Symp.on Iron Nutr. And Interact. Oral pres. Conf. Proc. 15 Tokyo Römheld (1986): *Physiol. Plant.* 70, 231-234.

Kaushik, K. K. 1997. Sustainable agriculture: issues and policy implications. *Productivity* 37(4): 142-147.

Khiari, L. and L. E. Parent. 2005. Phosphorus transformations in acid light-textured soils treated with dry swine manure. *Can. J. Soil Sci.* 85:75-87.

Kim, K. Y., D. Jordan D. and G. A. McDonald. 1997. Solubilisation of hydroxyapatite by *Enterobacter agglomerans* and cloned *Escherichia coli* in culture medium, *Biol. Fert. Soils* 24:347-352.

Klair, S., Adams, Z., Hider, R.C. and Leigh, R.A. 1995.The role of phytosiderophores in iron uptake by cereal plants. *J. Inorganic Biochem*. 59:109-109.

Kpomblekou, K. and M. A. Tabatabai. 1994. Effect of organic acids on release of phosphorus from phosphate rocks. *Soil Sci.* 158:442-453.

Lal, R. 2000. Controlling green house gases and feeding the globe through soil management. University Distinguished Lecture, Ohio State Univ., Columbus, February 17.

Lampkin, N., 1994. Organic farming: sustainable agriculture in practice. In: Lampkin, N., Padel, S.(Eds.), The Economics of Organic Farming. An International Perspective. CABI, Oxford.

Lampkin, N., Measures, M., 1995. 1995/96 Organic Farm Management Handbook University of Wales Elm, Farm Research Centre, Aberystwyth.

Lanting, H. 2007. Building a farmers owned company (Chetana) producing and trading fair trade–Organic products. Proceedings of the National Workshop on New Paradigm for Rainfed Farming, WASSAN, New Delhi, India.

Lazarovits, G., 2001. Management of soilborne plant pathogens with organic amendments: a disease control strategy salvaged from the past. *Can. J. Plant Pathol.*23: 1–7.

LEAF, 1991. Linking Environment and Farming: an Integrated Crop Management Project. LEAF publications, Stoneleigh, UK.

Lévai, L. 2004. XII. Int. Symp. on Iron Nutr. And Interact. Oral pres. *Conf. Proc. 15*, Tokyo.

Lovelock, C.E., Wright, S.F., Clark, D.A., Ruess, R.W., 2004. Soil stocks of glomalin produced by arbuscular mycorrhizal fungi across a tropical rain forest landscape. *J. Ecol.* **92**, 278–287.

McInerney, J., 1978. The technology of rural development (World Bank STA€ Working Paper No. 295).World Bank, Washington DC, USA.

Maity, T.K. and Tripathy, P. 2005.Organic Farming of Vegetables in India: Problems and Prospects. *Current Science* **88** : 262-274.

Makumba, W., Janssen, B., Oenema, O. and Akinnifesi, F.K. 2006.Influence of application of *Gliricidia* prunings as a source of N on the performance for maize. *Experimental Agriculture* **42**(1): 51–63

Marangoni, B., Toselli, M., Venturi, A., Fontana, M. and Scudellari, D. 2001. Effects of vineyard soil management and fertilization on grape diseases and wine quality. *IOBC/WPRS Bulletin* 243, 53-358.

Marwaha, B.C. and Jat, S.L. 2004. Statistics and scope of organic farming in India.*Fertilizer News* **49**(11): 41-48.

Mehmood, Y., Anjum, M. B., Sabir, M., Arshad, M. 2011: Benefit cost ratio of organic and inorganic wheat production: a case study of district Sheikhupura. *World Appl. Sci. J.*, 13, pp. 175-180.

Mueller, S., 1998.Evaluating the Sustainability of Agriculture.GTZ, Eschborn, Germany.

Mitchell, A.E. and Chassy, A.W. 2005.Antioxidants and the nutritional quality of organic agriculture.http://mitchell.ucdavis.edu/Is%20Organic%20Better.pdf

Nishiwaki, K. and Noue, T.I. 1996. The effect of animal manure compost applications on reducing the leaching of soil nutrients from mineral soils of vegetable upland field.

NPOP, 2003.National programme for Organic Production containing the standards for the organic products.Department of Commerce, Ministry of Commerce, Government of India.

NPOP, 2007.National programme for Organic Production containing the standards for the organic products.Department of Commerce, Ministry of Commerce, Government of India.

Olsson, P.A., Thingstrup, I., Jakobsen, I., Baath, E., 1999. Estimation of the biomass of arbuscular mycorrhizal fungi in a linseed field. *Soil Biol. Biochem.* 31, 1879–1887.

Organic Farming 2004, Indian Society of Soil Science, New Delhi.

Panneer, Selvam and Bheemaiah, G. 2005. Effect of integrated nutrient management practices on productivity, nutrient up take and economics of sunflower intercropped with *Azadirachta indica* and *Melia azadirach* trees. *Journal of Oilseeds Research* 22(1): 231–234.

Prasad, R. 2003. Protein energy malnutrition and fertilizer use in India.*Fertilizer News* 48(8):13-26.

Prasad, R. 2005. Organic farming vis-à-vis modern agriculture.*Current Science* 89:252-254.

Prasad, R. 2007. Modern Concepts of organic farming.Division of Agronomy, Indian Agricultural Research Institute, New Delhi, India.

Pretty, J., Hine, R., 2001: Reducing food poverty with sustainable agriculture: A summary of new evidence.(Last accessed on September 2011.

Rahman, M.M., Takahisa Amano, Inoue, H. and Matsumoto, Y. 2004. Nitrogen accumulation and recovery from legumes and N fertilizer in rice-based cropping systems. Forth International Crop Science Congress, Brisbane, Australia 26 September – 4 October 2004.

Rahudkar, W. B. and Phate, P. B. 1992. Organic farming: experiences of farmers in Maharashtra. Proceedings of National Seminar on Natural Farming, Rajasthan College of Agriculture, Udaipur, India.

Raj, Asha.K.and Geetha Kumari, V.L. 2001. Effect of organic manure and Azospirilluminoculation on yield and quality of Okra (*Abelmoschus esculentus* L).*Veg. Sci.*, 28(2): 179-181.

Ramesh, P., Singh, M., Subbha Rao, A. 2005: Organic farming: Its relevance to the Indian context. *Curr. Sci.*: 88, pp. 561–568.

Ramesh, P., Panwar, N. R., Singh, A. B., Ramana, S., Yadav, S. K., Shrivastava, R., SubbaRao, A., 2010: Status of organic farming in India. *Curr. Sci.* 98, pp. 1090–1194.

Randhawa, M.S. 1983. A History of Agriculture in India. Vol. III (1757-1947). Indian Council of Agricultural Research, New Delhi.pp. 422.

Rao, V.S. 1983. Principles of Weed Science, Oxford & IBH, New Delhi.

Research Bulletin of the Aichiken Agricultural Research Centre, No.28: 171-176 Nahas, E. 1996. Factors determining rock phosphate solubilisation by microorganism isolated from soil. *World J. Microb.Biotechnol.***12**:18-23.

Rosado-May, F.J., Werner, M.R., Gliessman, S.R. and Webb, R. 2005.Incidence of strawberry root fungi in conventional and organic production systems. *Applied Soil Ecology* **1**: 261–267.

Roychowdhury, R., Banerjee, U., Sofkova, S. and Tah, J. 2013. Organic farming for crop improvement and sustainable agriculture in the era of climate change. OnLine J. Biol. Sci. **13**(2): 50-65. doi: 10.3844/ojbsci.2013.50.65

Rigby, D., Howlett, D., Woodhouse, P., 1999. A Review of Indicators of Agricultural and Rural Livelihood Sustainability.Working Paper 1 in the series 'Sustainability Indicators for Natural Resource Management & Policy'.IDPM, University of Manchester.

Rigby, D., Young, T. and Burton, M., 2000. Why do farmers opt in or opt out of organic production? A review of the evidence. Symposium paper presented at the 2000 Agricultural Economics Society Conference, Manchester.

Sabhashini ,H.D., Malarvannan S. and Kumar P.2007. Effect of biofertilizers on yield of rice cultivars in Ponducherry, India. *Asian Journal of Agriculture Research* 1(3): 146-150.

Sagoe, C. I., T. Ando, K. Kouno and T. Nagaoka. 1998. Relative importance of protonsand solution calcium concentration in phosphate rock dissolution by organic acids.*Soil Sci. Plant Nutrition.* 44:617-625

Sakonnakhon, S.P.N., Toomsan, B., Cadisch, G., Baggs, E.M.,Vityakon, P., Limpinuntana, V., Jogloy, S. and Patanothai, A.2005. Dry season groundnut stover management practices determine nitrogen cycling efficiency and subsequent maize yields. *Plant and Soil* **272** (1/2): 183–199.

Sanghi, N. K. 2007. Beyond certified organic farming: an emerging paradigm for rainfed agriculture. Proceedings of the National Workshop on New Paradigm for Rainfed Farming: Redesigning Support Systems and Incentives, 27-29 September, IARI, New Delhi, India.

Sankaram, A. 2001. Organic farming: eco-technological focus for stability and sustainability. *Indian Farming* **9**: 7-11.

Save, B. H. 1992. Natural farming - an experience.Proceedings of National Seminar on Natural Farming, Rajasthan College of Agriculture, Udaipur, India.

Save, B. H. and Sanghavi, A. V. 1991. Economic viability of sustainable agriculture. Spice India, pp. 04.

Schliemann, G.K., Terblanche, J.H. and De Koch, I.S. 1983. Preparation and cultivationof plum soil. *Deciduous Fruit Grower* 1:20-27.

Sharma, A. K. 2001. A handbook of organic farming, Agrobios, Jodhpur, Rajasthan, India.

Sharma, PD, 2003, Prospects of Organic Farming in India, in Proceedings of National Seminar on Organic Products and Their Future Prospects, Sher-e-Kashmir University of Agricultural Sciences and Technology, Srinagar, pp 21-29.

Sharma, Kuldeep and Pradhan, Sudhir, 2011. Organic Farming: Problems and Prospects. *Yojana* 55: 68-70.

Sharma, R. K., Agrawal, M., Marshall, F. M. 2009: Heavy metals in vegetables collected from production and market sites of a tropical urban area of India. *Food Chem. Toxicol*. 47, pp. 583–591.

Sharma, R.C. and Sharma, H.C. 1990. Fertilizer phosphorus and potassium equivalents of somegreen manures for potatoes in alluvial soils of Punjab.*Trop.Agric*. 67:74-76

Sharma, R.C.; Govinda Krishnan, P.M.; Singh, R.P. and Sharma, H.C. 1988. Effect of FYM and green manures on crop yields and nitrogen needs of potato based cropping systems in Punjab.*J.Agric.Sci*;Camb. 110:499-564

Singh, A.P., Tripathi, R.S. and Mittra, B.N. 1999. Integrated nitrogen management in rainfed rice-linseed cropping system.*Tropical Agricultural Research and Extension* 2(2): 83-86.

Singh, U., Giller, K.E., Palm, C.A., Ladha, J.K. Breman, H. 2001. Synchronising N release from organic residues: opportunities for integrated management of N. 2nd International Nitrogen Conference, held during 14-18 October 2001 at Potomac, Maryland, USA.

Singh, Y.V and Dabas, J.P.S. 2012.Hormonize Organic Farming and Food Security. Kurkshetra, pp.29-34.

Som, M.G.; Hashim, H; Mandal, A.K. and Maity, T.K. 1992. Influence of organic manures on growth and yield of brinjal (*Solanum melongena* L). *Crop Research*, 5(1): 80-84.

Stockdale, E.A., Lampkin, N.H., Hovi, M., Keatinge, R., Lennartsson, E.K.M., Macdonald, D.W., Padel, S., Tattersall, F.H., Wolfe, M.S. and Watson, C.A. 2001. Agronomic and environmental implications of organic farming system. *Advances in Agronomy*.70: 261-327.

Stolze, M., Piorr, A., HaÈ ring, A., Dabbert, S. (2000) Environmental impacts of organic farming in Europe. Organic Farming in Europe: Economics and Policy. Stuttgart-Hohenheim 2000. Department of Farm Economics, University of Hohenheim, Germany.

Subba Rao, I. V. (1999): Soil and environmental pollution – A threat to sustainable agriculture *J. Indian Soc. Soil Sci*. 47, pp. 611–633.

Subba Rao, A., Subhash Chand and Srivastava, S. 2002. Opportunities for integrated plant nutrient supply system for crop/cropping system in different agro-eco-regions. *Fertiliser News*. 42:75-95.

Subbaiah, S. V., Rama Prasad, A.S., Kumar, R.M. and Surekha, K.2006. Interaction of organic manures and granular size of inorganic NPK fertilizers on nutrient uptake and grain yieldof irrigated rice in vertisols.18th World Congress of Soil Science, held during 9-15 July, Philadelphia, Pennsylvania, USA.

Subhash Chand and Pabbi, Sunil 2005 Organic Farming-Arising Concept .In Souvenir of Agriculture Summit organised by Govt. of India and FICCI, Vigyan Bhavan, New Delhi,1-8.

Subhash,Chand, 2008.Integrated Nutrient Management For Sustaining Crop Productivity and Soil Health. International Book Distribution Co., Lukhnow.

Subhash, Chand, L.L. Somani, 2005. Exploring Possibilities of Improving the Yield of Mustard through integrated nutrient management, *Int. J. of Tropical Agric.* **23**: 177-182.

Subhiah, K. 1991. Studies on the effect of N & *Azospirillum* in Okra.*South Indian Hort.*39 (1):37-44.

Suhane, R. 2007. Vermicompost. Publication of Rajendra Agriculture University, Pusa, Bihar, pp. 88.

Surange, S., A. G. Wollum, N. Kumar and C. S. Nautiyal. 1995. Characterization of *Rhizobium* from root nodules of leguminous trees growing in alkaline soils. *Can. J. Microbiol.***43**:891-894.

Suresh, H., Kunnal, L. B. (2004): Economics of Organic farming of Rice. In. Veeresh, G. K.,(Eds.). Operational Methodologies and Package of Practices in Organic Farming, APOF, Bangalore, pp. 77-78.

Swaminathan, C., Swaminathan, V. and Vijaylakshmi, K. 2007. Panchagavyya – Boon to organic farming. International Book Distributing Co. Lucknow (UP).

Swete-Kelly, D., 1996. Development and evaluation of sustainable production systems for steeplands Dlessons for the South Pacific. In: Sustainable Land Management in the South Pacific. Network Document no. 19, IBSRAM.

Tagliavini, M., Tonon, G., Scandellari, F., Quiñones, A., Palmieri, S., Menarbin, G. Gioacchini, P. and Masia, A. 2007.Nutrient recycling during the decomposition of apple leale (*Malus domestica*) and mowed grasses in an orchard. *Agri. Ecosystem Environ.* **118**: pp.191-200.

Taylor, D., Mohamed, Z., Shamsudin, M., Mohayidin, X., Chiew, E., 1993. Creating a farmer sustainability index: a malaysian case study. *American Journal of Alternative Agriculture* 8, 175±184.

Tej, Pratap and Vaidya, V.S. 2006. Organic Farmers Speak On Economics and Beyond. International Competence Centre for Organic Agriculture (ICCOA), Bangalore, India.

Teviotdale, B.L. and Hendricks, L. 2003. Survey of mycoflora inhabiting almond fruit and leaves in conventionally and organically farmed orchards. *Acta Horticulturae* **373**: pp.177–183.

Tiwari, K.N. 2005. Challenges of meeting nutrient needs in organic farming. *Indian Journal of Fertilizers*, **1**: pp.41-48, 51-59.

Van Bruggen, A.H.C., Semenov, A.M., 2003. In search of biological indicators for soilhealth and disease suppression. *Applied Soil Ecology* **15**: pp.13–24.

Veeresh, G. K. 1999. Organic farming ecologically sound and economically sustainable. *Plant Horti. Tech.* 1(3): pp.456-562.

Weibel, F., Bickel, R., Leuthold, S. and Alföldi, T. 2000. Are organically grown apple tastier and healthier? A comparative field study using conventional and alternative methods to measure fruit quality. *Acta Hort.* 517:pp.417-426.

Weymes, E., 1990. The Market for Organic Foods: a Canada-Wide Survey. Faculty of Administration,University of Regina, Saskatchewan.

White, K.D. 1970. *Agricultural History*, **44**: pp.281-290.

Whitelaw, M. A. 2000. Growth promotion of plants inoculated with phosphate solubilizing fungi. *Adv. Agron.* 69: pp.99-151.

Winter, C.K. and Davis, S.F. 2006.Organic Food. *J. Food Sci.* **9**: pp.117-124.

Worthington.V, 2001.Nutrional quality of organic verses conventional fruits, vegetables and grains. *Journal Alternative and Complimentary Medicine* 2: pp.161-173.

Zhu, Y.G., Miller, R.M., 2003. Carbon cycling by arbuscular mycorrhizal fungi in soil—plant systems. *Trends Plant Sci.* **8**, pp.407–409.

Glossary

A

A horizon: The surface horizon of a mineral soil having maximum organic matter accumulation, biological activity and/or eluviation of materials such as iron and aluminum oxides and silicate clays.

Acid rain : Atmospheric precipitation with pH values less than about 5.6, the acidity being due to inorganic acids such as nitric and sulfuric that are formed when oxides of nitrogen and sulphur are emitted into the atmosphere.

Acid soil: A soil that is acid in reaction throughout the root zone. Practically, this means a soil having pH less than 6.6; precisely, a soil with a pH value less than 7.0. Such a soil has more hydrogen (H) than hydroxyl (OH) ions in the soil solution. Acid soils are grouped into five categories.

Extremely acid: pH below 4.5, Very strongly acid: 4.5-5.0

Strongly acid: 5.1-5.5 , Medium acid: 5.6-6.0, Slightly acid: 6.1-6.5.

Acid soils are reclaimed by addition of liming materials.

Actinomycetes: A group of microorganisms intermediate between the bacteria and the true fungi that usually produce a characteristic branched mycelium. Includes many, but not all, organisms belonging to the order of Actinomycetales.

Additive: A material added to fertilizer to improve its physical condition.

Adhesion: It is process of molecular attraction that holds the surfaces of two substances (e.g. water and sand particles) in contact.

Adsorption complex: The group of organic and inorganic substances in soil capable of adsorbing ions and molecules.

Adsorption: It is a process of the attraction of ions or compounds to the surface of a solid. Soil colloids adsorb large amounts of ions and water.

Aerate: To impregnate with gas, usually air.

Aeration, soil: The process by which air in the soil is replaced by air from the atmosphere. In a well-aerated soil, the soil air is similar in composition to the atmosphere above the soil. Poorly aerated soils usually contain more carbon dioxide and correspondingly less oxygen than the atmosphere above the soil.

Aerobic : 1) Having molecular oxygen as a part of the environment 2) Growing only in the presence of molecular oxygen, as aerobic organisms 3) Occurring only in the presence of molecular oxygen (said of certain chemical or biochemical processes, such as aerobic decomposition).

Agriculture: the science, art, or practice of cultivating the soil, producing crops, and raising livestock and in varying degrees the preparation and marketing of the resulting products.

Agroecology: is a whole systems approach to agriculture that applies ecology to the design and management of sustainable production systems.http:// agroecology.org/

Alternative farming and **alternative agriculture**: general terms that apply to agricultural production methods, agricultural enterprises, and/or crops that are different from traditional or conventional ones.

Aquaculture: the farming of aquatic-based species: fish, crustaceans, mollusks and plants.

Alkali Soil: A soil that contains sufficient sodium salt to interfere with the growth of most crop plants.

Alkaline soil: A soil that is alkaline in reaction throughout the root zone or for a major part of the root zone. Precisely, any soil having a pH value greater than 7.0. Practically, a soil having a pH above 7.3 is called alkaline. An alkaline soil is reclaimed by addition of gypsum or sulphur. Soils having pH from 7 to 7.5 are sometimes referred to as alkali soils. They correspond to black alkali soils and occur in irregular patches. They can be reclaimed by adding calcium or magnesium.

Amendment soil: Any substance other than fertilizers, such as lime, sulphur, gypsum, and sawdust, used to alter the chemical or physical properties of a soil, generally to make it more productive.

Amino acids: Nitrogen-containing organic acids that couple together to form proteins. Each acid molecule contains one or more amino groups ($-NH_2$) and at least one carboxyl group (-COOH). In addition, some amino acids contain sulphur.

Ammonification: It is biochemical process whereby ammoniacal nitrogen is released from nitrogen-containing organic compounds.

Ammonium fixation: The entrapment of ammonium ions by the mineral or organic fractions of the soil in forms that are insoluble in water and at least temporarily nonexchangeable.

Arid climate: Climate in regions that lack sufficient moisture for crop production without irrigation. In cool regions annual precipitation is usually less than 25 cm. It may be as high as 50 cm in tropical regions. Natural vegetation is desert shrubs.

Available nutrients in soils: A part of the plant nutrient in the soil that can be taken up by growing plants immediately. Available nitrogen is defined as the water soluble nitrogen plus the part that can be made soluble or converted into free ammonia. Available phosphoric acid is that part which is soluble in water or in a week dilute acid such as 2% citric acid. Available potash is defined as that portion which is soluble in water or in a solution of ammonium oxalate.

Autotrophs : Group of micro organisms which obtain their energy from sunlight or by oxidation of inorganic compounds. If the energy is derived from sunlight, the group is called photo-autotroph. The other group which derives energy from oxidation of inorganic material is called chemo-autotroph. Algae are photo-autotrophs. Some autotroph bacteria which play an important part in availability of plant nutrients are:

1. *Nitrosomonas* – oxidize ammonium to nitrate

2. *Thiobacillus* – oxidize inorganic S to sulphate

3. *Nitrobacters* – oxidize nitrite to nitrate.

4. *T.ferroxidans*-oxidise ferrous iron to ferric form.

Azofication: Non-symbiotic nitrogen fixation by the *azotobacter* group of soil bacteria which use organic matter as a source of energy and are able to obtain nitrogen from the atmosphere to build up their body protein. After the death of the bacterial cell, this nitrogen is returned to the soil for use by higher plants.

Azolla: A small floating aquatic fern that is found in the tropics and sub-tropics and some temperate zones of the world. It is a member of order Salviniales and comprises of seven recognized living species, namely :

1. *Careliniana*   5. *Nilotica*

2. *Filiculoides*   6. *Pinnata*

3. *Mexicana*   7. *Rubra*

4. *Microphylla*

For rice cultivation azolla is applied as green manure both by basal application and as top dressing. Under optimum conditions an azola crop may produce 3 kg or more N/ha/day. Thus a 20 day old azolla crop can produce 60 kg N/ha. Two basal crops of 20 days each thus can produce almost 120 kg N/ha for the following rice crop.

Azotobacter: Free living bacteria capable of utilizing atmospheric nitrogen and fixing it for their biosynthetic reaction. These can also utilize ammonium, nitrate, nitrite, urea and sometimes organic nitrogen.Azotobacter are essentially aerobic. There are five species of azotobacter recognized on the basis of cell shape, pigmentation and mobility :

1. *Azotobacter chroococcu*

2. *Azotobacter beijerinckii*

3. *Azotobacter vinelandii*

4. *Azotobacter macrocytogenes*

5. *Azotobacter agilis.*

B

Bacterial culture: Any media enriched with any particular bacteria. The culture describes the bacteria contained in it e.g.Rhizobium culture: is used for inoculating seeds of leguminous crops.

Biodiversity: (generally refers to) the number of varieties or species of life in a given ecosystem.

Biogas: The gas produced from organic waste by microbiological reaction under anaerobic conditions. The organic wastes used in the production of biogas are generally cattle yard waste, human waste, vegetative crop residues, unwanted aquatic plants and weeds. Biogas is sometimes termed after the material or method of gas production, e.g. marsh gas, sewage gas, sludge gas, digestor gas and *gobar* (cowdung) gas, etc. As a fuel, biogas is a potential source of energy. It is usually have the following composing.

Methane = 50-60%, Carbondioxide = 30-40%, Hydrogen = 5-10%

Hydrogensulphide = traces, Water vapour = traces

Biological test: A technique for assessing nutrient status of soil using biological material. The following types of tests are commonly used:

1. Field test: Involves use of field crops in experimental or farmers' fields.

    2. Laboratory or green house tests:Mitscherlich pot culture

    3. Lettuce pot culture

    4. Neubauer seedling method

    5. Sunflower pot culture

    6. Microbiological tests :

    7. Azotobacter culture

    8. Sackett and Stewart technique

    9. Aspergillus Niger test

    10. Mehlich's Cunninghamella Plaque method.

Biomass: The amount of living matter in a given area.

Biodynamic agriculture is a concept of agriculture that sees the farm as a living, dynamic, spiritual entity and attempts to bring it into balance. The Demeter Association establishes the specific guidelines for Biodynamic production and certification. While generally regarded as a type of organic farming system, Biodynamic agriculture is considerably more rigorous. http://www.biodynamics.com/http://www.demeter-usa.org/

Biuret ($C_2O_2N_3H_5$): A chemical compound formed by the combination of two molecules of urea with the release of a molecule of ammonia, when the temperature during the urea manufacturing process exceeds a certain level. Fertiliser grade urea contains variable amounts of biuret. Biuret is toxic to plants, particularly when it is applied through sprays. As per the Indian fertilizer Legislation (FCO, 1985) the biuret content in urea should not exceed 1.5%.

Blood Meal: Dried blood obtained from a slaughter house. The blood lot is treated with copper sulphate at the rate of 2 oz per 100 lb of blood lot. It is then heated to evaporate from moisture and then dried under the sun. After drying it is powered, bagged and sold as blood meal. Various other treatments like testing blood serum or clots with lime or very dilute hydrochloric acid or absorbing the blood in organic wastes and then drying in the sun, are also used in the preparation of blood meal. It is quite good organic fertilizer containing 10 to 12% nitrogen and 1-2% P_2O_3.

Blue-Green Algae: Blue-green algae belong to the class Cyanophyceae or Myxophyceae. These are blue-green in colour because of the presence of pigments like chlorophyll, carotenes, xanthophylls, physcocyanin and phycoerythrin. Blue green algae are divided into five orders:

   1. *Chroococcales,* 2 *Chmaesiphonates,* 3 *Plerocapsales*

   4 *Nostocales,* 5 *Stigonematales.*

Important genera concerned with nitrogen-fixing activity are: Anabaena, Nostoc, and Cylindrospermum. According to conservative estimates, blue-green algae can fix about 25-30 kg N/ha per cropping season.

The algae also give some additional benefits.

1. Because of photosynthetic activity they release oxygen which is a key factor in rice culture.

2. They increase the organic content of soil.

3. They improve the physical structure by increasing water stable aggregates.They supply growth promoting substances like Vitamin B_{12}, IAA-IPA, Anthramilic acid and other allied compound.

Bonemeal, raw: A fertilizer made of dried animal bones finely ground. It contains 20 to 24 per cent P_2O_5. The availability of its plant food depends largely upon how fine it is ground.

Bonemeal, steamed: A product made from grinding bones previously treated with steam under pressure. It contains one to two per cent nitrogen and 22 per cent phosphorus.

Bulky organic manures: These manures are bulky in nature and supply plant nutrients in small quantities and organic matter in large quantities.

Biological control (often shortened to **bio-control**) is the practice of using living organisms (e.g. predators, parasites, or diseases) to control or manage a separate, harmful organism (*e.g.* weeds, plant pathogens, or insect pests.) For a good source of bio-control information, refer to the North Carolina Biological Control Information Center Web site: http://cipm.ncsu.edu/ent/biocontrol/

Biological farming is the term often used in Europe to refer to organic farming.

Biotechnology refers to the integration of the biological sciences and technology. Biotechnology can include anything from selective plant and animal breeding to gene manipulation and cloning. This term can also be applied to the use of biological organisms in industry (*e.g.* fermentation processes used to make foods; manufacturing pharmaceuticals from microorganisms, etc.).

Biointensive agriculture is an organic agricultural system that is practiced on a relatively small scale. It is also referred to as biointensive mini-farming or biointensive gardening. This system focuses both on obtaining maximum yields from a minimal area of land and on long-term sustainability in a closed system. http://www.growbiointensive.org/

Biointensive IPM is an attempt to take integrated pest management back to its ecological roots. It emphasizes many of the concepts inherent in IPM, such as understanding pest biology, rotating crops to disrupt pest life cycles, and using resistant varieties. However, only reduced risk pesticides are used, and then only as a last resort after other preventative tactics have proved ineffective. For more information, refer to the ATTRA publication on this topic:http://attra.ncat.org/attra-pub/ipm.html

C

Carbon: An essential plant element. Plants take their carbon requirement from atmospheric carbon dioxide. Since it is never deficient in the atmosphere to become a limiting factor in growth, it is not supplied in any form as fertilizer. Carbon present in organic compounds in soil cannot be used as a nutrient source of carbon.

Carbon cycle: The sequence of transformations whereby carbon dioxide is fixed in living organisms by photosynthesis or by chemosynthesis, liberated by respiration and by the death and decomposition of the fixing organism, used by heterotrophic species, and ultimately returned to its original state.

Carbon-nitrogen ratio: The ratio of the weight of total organic carbon to the weight of total nitrogen in a soil or in an organic material. C:N ratio of wheat straw is nearly 80:1 while that of soil is 10:1. When undecomposed straw with a high C:N ratio is applied to the soil, its C:N ratio is reduced through bacterial decomposition. To speed up bacterial decomposition, nitrogenous fertilizer is added to the soil at the time of turning under straw.

Capillary water: The water held in the "capillary" or *small* pores of a soil, usually with a tension >60 cm of water.

Chelates: An organic compound capable of holding the plant nutrient in a form which prevents it from getting tied with other elements in the soil, thus keeping it more or less in available form for the plant. The term refers to the claws of a crab illustrative of the way in which the atom is held. For examples EDTA, DTPA, HEEDTA and CDTA.

Chlorosis : A condition in plants relating to the failure of chlorophyll (the green colouring matter) to develop. Chlorotic leaves range from light green through yellow to almost white.

Certified Naturally Grown (CNG) is a grassroots alternative to the USDA organic certification programme and is primarily aimed at small farmers. This organization uses NOP regulations as the basis for its own high standards of organic production. However, the CNG puts less emphasis on record-keeping and encourages the sharing of advice between inspectors and farmers. Farms that have been inspected and certified by CNG members can use the Certified Naturally Grown logo on their products. http://www.naturallygrown.org/

Conservation tillage refers to a broad range of tillage practices that leave 30% or more of the crop residue from the previous crop on a field when planting the next crop. No-till, strip-till, ridge-till, and mulch-till are types of conservation tillage. Benefits include reducing soil erosion and minimising disturbances to the soil ecosystem.

Conventional agriculture or **conventional farming** refers to traditional agricultural practices, such as a reliance on pesticides and synthetic fertilizers. These terms are often used in contrast to organic or sustainable agricultural systems. Growers who are not "organic" are referred to as "conventional" farmers.

Clay humus complex : It is a complex formed due to interaction of the clay and humus colloids through different mechanisms and force like, cation bridge, hydrogen bonding, vander waals forces.

Colloid, soil (Greek, glue-like) Organic and inorganic matter with very small particle size and a correspondingly large surface area per unit of mass.

Compost : A mass of rotted organic matter made from waste.

Compost : Organic residues, or a mixture of organic residues and soil, that have been piled, moistened, and allowed to undergo biological decomposition. Mineral fertilizers are sometimes added. Often called "artificial manure" or "synthetic manure" if produced primarily from plant residues.

Composting: It is largely a biological process in which micro-organisms of both types namely aerobic and anaerobic, decompose the organic matter and lower the carbon-nitrogen ratio of the refuse. The final product of composting is a well-rooted manure known as compost.

Conditioner (of fertilizer): A material added to a fertilizer to prevent caking and to keep is free-flowing.

Contour Farming: the practice of tilling sloped land along lines of consistent elevation in order to conserve rainwater and to reduce soil losses from surface erosion.

Covered Commodity: (also called **programme commodity**) Agricultural products for which Federal support programmes are available to producers.

Crop Insurance: the major USDA program that helps farmers manage risks of crop losses.Subsidized crop insurance remains the primary form of assistance provided by the Federal Government against bad weather, plant diseases, and other natural hazards; disaster assistance payments are also frequently provided.

Crop Rotation: the successive movement of crops from one area to another as a means to improve soil fertility, to reduce disease, and to reduce insect populations.

Conduction: The transfer of heat by physical contact between two or more objects.

Cotton seed cake: A by-product of the cotton seed crushing industry, when it is undecorticated, it is non-edible containing 3.9 per cent nitrogen, 18 1.8 per cent phosphoric acid and 1.6 per cent potash. When decorticated, cotton seed cake becomes edible oilcake containing higher percentages of plant nutrients. Decorticated cotton seed cake contains 6.4 per cent nitrogen, 2.9 per cent phosphoric acid and 2.2 per cent potash.

Crop rotation : A planned sequence of crops growing in a regularly recurring succession on the same area of land, as contrasted to continuous culture of one crop or growing different crops in haphazard order.

D

Deficiency: A condition of insufficient supply of essential plant nutrients required for various metabolic functions.Plants cannot reach optimum growth, if any one of the essential elements is limited in supply. Deficiency may be caused due to any one of the following reasons:

The nutrient is in short supply in soil.

The nutrient is there in soil, but is not in the available form in which plants can absorb it.

There is the antagonistic effect of other elements. See also Antagonistic effect. This is called induced deficiency as the nutrient in question is not being metabolised because of the excess of other antagonistic elements.

Denitrification: The process by which nitrates or nitrites in the soil or in farmyard manure are reduced to ammonia or free nitrogen by bacterial action. The process results in the escape of nitrogen into the air and is therefore wasteful.

Desalinization: Removal of salts from saline soil, usually be leaching.

Diatomaceous earth: A geologic deposit of fine, grayish, siliceous material composed chiefly or wholly of the remains of diatoms. It may occur as a powder or as a porous, rigid material.

Diatoms : Algae having siliceous cell walls that persist as a skeleton after death, any of the microscopic unicellular or colonial algae constituting the class Bacillariaceae. They occur abundantly in fresh and salt waters and their remains are widely distributed in soils.

Diffusion : The transport of matter as a result of the movement of the constituent particles. The intermingling of two gases or liquids in contact with each other takes place by diffusion..

Dry farming : The system in which field crops are raised with an annual rainfall of less than 25 inches, without irrigation facilities. From plant nutrition point of view, lesser doses of fertilizers are recommended to dry farming areas for all field crops, compared to irrigated farming.

Dryland farming : The practice of crop production in low rainfall areas without irrigation.

Duff : The matted, partly decomposed organic surface layer of forest soils.

E

Ecosystem: a system that includes all living organisms in an area as well as its physical environment functioning together as a unit.

Eutrofication: Pollution with unwanted nutrients.

Essential plant nutrient: A nutrient essential for plants for proper growth and development. At present, there are sixteen essential plant nutrients recognized by plant physiologists. These are carbon, hydrogen, oxygen, nitrogen, phosphorus, calcium, potassium, magnesium, sulphur, manganese, boron, copper, zinc, iron, molybdenum and chlorine.

Ecological farming is a term used in Europe to refer to organic agriculture, but with a greater emphasis on environmental concerns.

Environmentally friendly is a general term used to describe products or services that have resulted in minimal to no harm to the environment.

Ethnic market refers to consumers or buyers that share a common cultural or racial background.

Eucaryotic : In biology, referring to the type of cells with a distinct nucleus with a nuclear membrane, characteristic of fungi, protozoa, algae, plants and animals.

Evapotranspiration : Evaporation plus transpiration.

Extracellular : Outside the cell. Extracellular enzymes are excreted by some bacteria and fungi.

F

Facultative organism : An organism capable of both aerobic and anaerobic metabolism.

Fallow : Cropland left idle in order to restore productivity, mainly through accumulation of water, nutrients, or both. Summer fallow is a common stage before cereal grain in regions of limited rainfall. The soil is kept free of weeds and other vegetation, thereby conserving nutrients and water for the next years crop.

Fallowing : Keeping the land free of a crop or weeds for a period of time. This is done to restore soil productivity mainly through accumulation of water, nutrients or both. Fallow during the kharif season is a common practice in wheat-growing regions which receive limited rainfall. During fallow, the field is cultivated to control weeds and to help the storage of moisture for the succeeding crop.

Farmyard manure (FYM): It is a product of decomposition of the liquid and solid excreta of animals stored in the farm. In western countries, straw or other litter used as bedding is also included along with the animal excreta. In India, since straw is used mainly for fodder purposes, farmyard manure is made mainly from animal excreta. The composition of FYM varies with the nutrient content in excreta and with the method of preparation. On an average it contains 0.5% N, 0.2% P_2O_5 and 0.5% K_2O.

Fauna : The animal life of a region.

Fermentation : A set of metabolic processes by which anaerobic organisms obtain energy by converting sugars to alcohols or acids and CO_2.

Fertility ofsoil : The quality of a soil that enables it to provide essential chemical elements in quantities and proportions for the growth of specified plants.

Fertilizer: Any natural or manufactured material, dry or liquid, added to the soil in order to supply one or more plant nutrients. The term is generally applied to commercially manufactured materials other than lime or gypsum. When fertilizers are sold on a large scale, they are called commercial fertilizers.

Field capacity (field moisture capacity) **:** The percentage of water remaining in a soil two or three days after its harving been saturated and after free drainage has practically ceased.

Field experiments: Experiments conducted in the field to determine the type and amount of fertilizers to suit particular soil type and crop.

Fixation (in soil)**:** Conversion of a soluble material such as a plant nutrient like phosphorus, from a soluble or exchangeable form to a relatively insoluble form. To reduce fixation of phosphate, phosphatic fertilizers are brought in lesser contact with the soil particles and applied closer to the plant root through band placement..

Fortification: The process of putting an additional quantity of fertilizer to increase the nutrient content of manure/ compost, e.g. in the ADCO process of compost-making superphosphate is added to fortify the phosphate content of the manure.

G

Green refers to environmentally friendly products that are derived from recycled materials or renewable resources.

Holistic Management :is a sustainable farm planning tool that views the farm, the family, and the community as a whole, rather than as separate entities. Holistic management emphasizes the establishment of long-range goals, while also meeting immediate needs. Other related concepts include Whole Farm Planning, comprehensive farm planning, and integrated farm management. http://www.holisticmanagement.org/

Integrated Pest Management (IPM) is a pest management strategy that uses a combination of biological, cultural, and chemical tools to reduce crop damage from insects, diseases, and weeds.

These strategies are employed in such a way as to minimize environmental risks, economic costs, and health hazards. IPM uses such techniques as monitoring pest populations, observing weather conditions, understanding pest cycles, biological control, and crop rotation to reduce and/or manage pest populations. Visit the University of Kentucky IPM Web site for more information and resources: http://www.uky.edu/Ag/IPM/ipm.htm

Irradiation is a method of disinfesting, sterilising and/or preserving food using ionising radiation.

Greenleaf manuring : This refers to turning under of green leaves and tender green twigs collected from shrubs and trees grown on bunds, waste lands and nearby forest areas.The common shrubs and trees useful for this purpose are glyricidia (*Glyricidia maculate*), *Sesbania speciosa*, karanj (*Pongamia pinnata*) etc.

Green manure crop: Any crop grown and buried into the soil for improving the soil condition by addition of organic matter. Such crops are legumes and non-legumes, but mostly legumes e.g. sannhamp (*Crotolaria juncia*), Dhaincha – (*Sesbania aculeata*)

Green manuring: A practice of ploughing or turning into the soil undecomposed green plant material for improving the physical condition of the soil or for adding nitrogen when the green manure crop is legume. Two types of green manuring are being practiced by the cultivators of India. These are i) green manuring *in situ* and ii) Greenleaf manuring.

Green manuring *in situ*: A practice in which green manure crops are grown and buried in the same field which is to be green-manured. The crops are grown alone or intercropped with the main crop. Important green manure crops used in this system are sannhemp (*Crotalaria juncea*), dhaincha (*Seshania aculeate*), pillipesara (*Phaseolus trilobus*), and guar (*Cyamopsis psoraloides* or *tetragonoloba*).

Greenhouse effect: The warming of the earth's surface and atmosphere owing to absorption of out-going radiation by CO_2, CH_4 and H_2O (like absorption by glass).

Groundnut cake: A by-product of the oil industry. This is an edible oil cake. It contains 7.3 per cent nitrogen, 1.5 per cent phosphoric acid and 1.3 per cent potash.

Guano : In some parts of America used synonymously with fertilizer. It is derived from the Spanish word meaning dung. Guano is made up of excrement of seafowl, together with their body remains. Deposits of guano found in many islands may vary from a few inches to over 200 feet in thickness. Peru is one of the pioneer producers and exporters of guano. It is also obtained from excrement and dead remains of sea creatures other than seafowl. It is also called bat guano, seal guano, fish guano, whale guano, etc. Some guanos produced from sheep and goat herds are called sheep guano or goat guano. The colour of guano may vary from grey to dark brown. The chemical composition of guano also varies. Its nitrogen content may vary from 4 to 16% and total P_2O_5 may range between 12 and 26%. When nitrogen content is very high, say, 8-16%, it is termed as nitrogenous guano. When phosphorus content is very high, say, 20-25% P_2O_5, it is termed as phosphatic guano. The potash (K_2O) content of guano is 2 to 3%.

The product obtained after treatment of guano with sulphuric acid is called dissolved guano. The available phosphorus and nitrogen content increase as a result of this treatment. Some times any one or all of the three nutrients, namely, N, P_2O_5 and K_2O are added to guano to make it a balanced fertilizer. This is called rectified or fortified guano.

H

Halophyte : A plant that requires or tolerates a saline (high salt) environment.

Hemicellulose : A group of complex carbohydrates; polysaccharides that, unlike starch and cellulose, contain other sugars besides glucose. Important to plant cell walls.

Herbicide : A chemical that kills plants or inhibits their growth; intended for weed control.

Herbivore : A plant-eating animal.

Heterotroph : An organism capable of deriving energy for life processes only from the decomposition of organic compounds and incapable of using inorganic compounds as sole sources of energy or for organic synthesis.

Horn-and-hoof-meal: A manure prepared from horns and hoofs of animals. Such a meal contains about 13 per cent nitrogen.

Horticulture: The art and science of growing fruits, vegetables, and ornamental plants.

Humic acid: A mixture of variable or indefinite composition of dark organic substances, precipitated upon acidification of a dilute alkali extract from soil.

Humid climate: Climate in regions where moisture, when distributed normally throughout the year, should not limit crop production. In cool climate annual precipitation may be as little as 25 cm; in hot climates, 150 cm or even more. Natural vegetation in uncultivated areas is forests.

Humification: The process involved in the decomposition of organic matter and leading to the formation of humus.

Humin : The fraction of the soil organic matter that is not dissolved upon extraction of the soil with dilute alkali.

Humus : That more or less stable fraction of the soil organic matter remaining after the major portions of added plant and animal residues have decomposed. Usually it is dark in colour. The term is often used synonymously with soil organic matter.

Hydration : The chemical combination of water with another substance.

Hydrologic cycle : The circuit of water movement from the atmosphere to the Earth and back to the atmosphere through various stages or processes, as precipitation, interception, runoff, infiltration, percolation, storage, evaporation, and transpiration.

Hydro-mulching : Technique of spraying a slurry of fiber, seed, fertilizer, and chemicals onto roadsides for erosion control.

Hydrous oxides, hydroxyoxides : Sesquioxides. Oxides of Fe, Al, or similar metals, with different proportions of water and hydroxyl in the structure.

Hybrid: the offspring of genetically dissimilar parents, cross-bred across breeds, subspecies,species, varieties, or genera.

I

Immobilization: The conversion of an element from the inorganic to the organic form in microbial tissues or in plant tissues, thus rendering the element not readily available to other organisms or to plants.

Inactivated organisms: In waste disposal, harmful organisms made harmless by reaction with the soil and soil organisms.

Inoculation : The process of introducing pure or mixed cultures of microorganisms into natural or artificial culture media.

Intercrop : Two or more crops grown together on the same piece of land at the same time.

Integrated plant nutrition system (IPNS) : It is a concept proposed by FAO where the basic goal is the maintenance or adjustment and possibly improvement of soil fertility and of plant nutrient supply to an optimum level for sustaining the desired crop productivity through optimization of the benefits from all possible sources of plant nutrients in an integrated manner.

Interveinal chlorosis : Chlorosis only between leaf veins..

L

Labile: Descriptive of a substance in soil that readily undergoes transformation or is readily available to plants.

Leaching : Removal of plant nutrients in solution by the passage of water through soil. This is one of the ways in which plant nutrients are lost from the soil. Among the major nutrients, nitrogen is lost in large quantities by leaching.

Legume Inoculation : Treatment of legume seed with rhizobium culture. There is a specific symbiotic relationship between different species of rhizobium and various legume crops. For example, a rhizobium specie that will live symbiotically with soyabean will not do so with alfalfa. It is, therefore, necessary to use only specific cultures for different crops.

Legume: A pod-bearing member of the leguninosae family, one of the most important and widely distributed plant families. Includes many valuable food and forage species, such as peas, beans, peanuts, clovers, alfalfas, sweet clovers, lespedezas, vetches, and kudzu. Nearly all legumes are associated with nitrogen-fixing organisms.

Lichen : Symbiosis between fungi and algae or bluegreen bacteria, commonly forming a flat, speading growth on surfaces of rocks and tree trunks.

Lignin : The complex organic constituent of woody fibers in plant tissue that, along with cellulose, cements the cells together and provides strength. Lignins resist microbial attack and after some modification may become part of the soil organic matter.

Livestock: (in the broadest sense) any animal raised on the farm for a profit, including but not limited to cattle, swine, sheep, poultry, fish, and bees.

M

Macronutrient: A nutrient required by plants in relatively large amounts. There are three macronutrients namely, nitrogen, phosphorus and potassium which are also called major nutrients or macro elements. Other nutrients which are normally required in lesser but still considerable amounts like calcium, magnesium and sulphur are called secondary nutrients. Some authors group secondary nutrients along with macronutrients.

Manure : The excreta of animals – dung and urine, with straw or other materials used as the absorbent. The decomposed manure is called farmyard manure or farm manure or barnyard manure. The average composition of well-rotted farmyard manure is 0.5 per cent nitrogen, 0.3 per cent P_2O_5 and 0.5 per cent K_2O.

Marsh : Periodically wet or continually flooded area with the surface not deeply submerged. Covered dominantly with sedges, cattails, rushes, or other hydrophytic plants. Subclasses include freshwater and saltwater marshes.

Methane, CH_4: An odersless, colourless gas commonly produced under anaerobic conditions. When released to the upper atmosphere, methane contributes to global warming.

Microfauna : That part of the animal population which consists of individuals too small to be clearly distinguished without the use of a microscope. Includes protozoans and nematodes.

Microflora : That part of the plant population which consists of individuals too small to be clearly distinguished without the use of a microscope. Includes actinomycetes, algae, bacteria, and fungi.

Micronutrients: The essential plant nutrients required in minute quantities. These nutrients are seven in number, namely, iron, manganese, boron, molybdenum, copper, zinc and chlorine. Micronutrients are also called 'minor elements' or 'trace elements'.

Micro-organisms : The most primitive plant and animal life whose structure is very simple. Their size being small, they are often found in soil and have direct or indirect bearing on soil formation and soil fertility. They are divided into two main groups:

> Micro flora, *e.g.* bacteria, actinomycetes, fungi and algae.

> Micro fauna, *e.g* protozoa.

Mineral soil : A soil consisting predominantly of, and having its properties determined predominantly by mineral matter. Usually contains <20% organic matter, but may contain an organic surface layer upto 30 cm thick.

Mineralization : The conversion of an element from an organic form to an inorganic state as a result of microbial decomposition.

Monoculture: the agricultural practice of growing a single crop across a wide area of land; crop system reliant on a narrow genetic mix.

Muck: Highly decomposed organic material in which the original plant parts are not recognisable. Contains more mineral matter and is usually darker in colour than peat.

Muck soil :(1) A soil containing 20-50% organic matter. (2) An organic soil in which the organic matter is well decomposed.

Mull : A humus-rich layer of forested soils consisting of mixed organic and mineral matter. A mull blends into the upper mineral .

Mycorrhiza : A mycorrhiza is an infected root system arising from the root lets of a seed plant. The word mycorrhiza (my-koe-rye-zee), derived from Greek meaning "fungus root" Mycorrhizae are fungi that form symbiotic association of a fungus with the roots of a higher plants.

N

Necrosis : Death associated with discoloration and dehydration of all or parts of plant organs, such as leaves.

Neem coated urea : Neem has a nitrification inhibition property, hence is used in India for coating of urea granules to produce slow release fertilizer. The technique as developed at the Indian Agriculture Research Institute for coating urea with neem cake is as follows: About 100 kg urea is mixed in a drum with a solution of 1 kg coaltar in 2 litres of kerosene. To this, 20 kg of powdered neem cake is added and thoroughly mixed.

Nematodes : Very small worms abundant in many soils and important because some of them attack and destroy plant roots.

Neutral soil : A soil in which the surface layer, at least to normal plow depth, is neither acid nor alkaline in reaction. In practice this means the soil is within the pH range of 6.6 – 7.3.

Night soil : Night soil is human excreta, solid and liquid, In India, it is directly applied to the soil to a limited extent but converted mainly as town compost. In cities which have sewage facilities, sewage water and sludge are used directly to raise crops. On an average night-soil contains 5.5 per cent nitrogen, 4.0 per cent phosphorus (P_2O_5) and 2.0 per cent potash (K_2O) on oven dry basis.

Nitrification : The formation of nitrates and nitrites from ammonia (or ammonium compounds) as in soils, by micro-organisms.

Nitrogen cyle : The sequence of chemical and biological changes undergone by nitrogen as it moves from the atmosphere into water, soil, and living organisms, and upon death of these organisms (plants and animals) is recycled through a part or all of the entire process.

Nitrogen fixation : The assimilation of free nitrogen from the soil air by soil micro-organisms and the formation of nitrogen compounds that eventually become available to plants. The nitrogen fixing organisms associated with legumes are called symbiotic; those not definitely associated with higher plants are non-symbiotic.

Nucleic acids : Complex compounds found in plant and animal cells may be combined with proteins as nucleoproteins.

Nutrient : Any mineral element that functions in plant, animal and other organism, metabolisms, whether or not its action is specific or a chemical that an organism needs to live and grow or a substance used in anorganism's metabolism, which must be taken in from its environment.

Nutrient deficiency symptoms: When any essential plant nutrient is seriously lacking in the soil, plants growing on it show certain colour development in leaves and certain changes in growth. These symptoms varies from crop to crop and with the degree of deficiency. On the basis of deficiency symptoms, fertilizer recommendations can be done.

Natural :refers to foods or food additives that are not produced or manufactured. In addition, natural foods contain no artificial ingredients, including preservatives, and have undergone minimal processing. The FDA does not regulate the use of this term except when used on meat and poultry. Natural is *not* synonymous with "organic" or "sustainable."

Natural farming is a highly refined method of working closely with nature to obtain high yields with little labour involvement. The founder, Japanese farmer Masanobu Fukuoka, refers to his method as "do-nothing farming."

Naturally grown is a general term that suggests the crop was produced without pesticides or other synthetic chemicals.

Niche market refers to a relatively narrow, but potentially profitable market that targets a specific group of buyers. These markets are characterized by a unique or differentiated product, a limited number of buyers and sellers, and the existence of potential barriers for entering the market.

Nutrient management relates to managing the amount, timing, form, and placement of soil amendments used in plant production. The current focus is on optimizing crop production and economic returns while also taking into consideration environmental concerns. For additional information, refer to Environmental Protection Agency Ag101 Web site: http://www.epa.gov/oecaagct/ag101/

O

O horizon : Organic horizon of mineral soils.

Oilcakes : When oil is extracted from oilseeds, the remaining solid portion is the oilcake. Oilcakes are of two types, edible oilcakes which can be safely fed to livestock, and non-edible oilcakes which are not fit for feeding to livestock. Oilcakes are added to the soil as concentrated organic manures. They supply organic matter and all the three major plant nutrients (N, P_2O_5, and K_2O) but mostly they supply nitrogen.

Organic colloids: In soils, these present as humus which is a temporary intermediate product left after considerable decomposition of plant and animal residues. Organic colloids having high surface area with higher charge density and higher cation exchange capacity.

Organic farming: According to the U.S. Department of Agriculture "a production system which avoids or largely excludes the use of synthetically compounded fertilizers, pesticides, growth regulators and livestock feed additives. To the maximum extent feasible, organic forming systems rely upon crop rotations, crop resideues, animal manures, legumes, green manures, off-farm organic waste, mechanical cultivation, mineral bearing rocks and aspects of biological pest control to maintain soil productivity and tilth, to supply plant nutrients and to control insects, weeds and other pests".

Organic fertilizer : By product from the processing of animal or vegetable substances that contain sufficient plant nutrients to be of value as fertilizers.

Organic soil : A soil that contains at least 20% organic matter (by weight) if the clay content is low and at least 30% if the clay content is as high as 60%.

Organic soil materials (As used in Soil taxonomy in the United States). (1) Saturated with water for prolonged periods unless artificially drained and having 18% or more organic carbon (by weight) if the mineral fraction is more than 60% clay, more than 12%, organic carbon if the mineral fraction has no clay, or between 12 and 18% carbon if the clay content of the mineral fraction is between 0 and 60%. (2) Never saturated with water for more than a few days and having more than 20% organic carbon. Histosols develop on these organic soil materials.

Organic matter: The terms organic means "living", therefore, organic matter is any material derived from living organism like plants and animals. The organic matter content in soil is an important indicator of nutrient status especially of nitrogen content in soils. Generally, the status of soil organic carbon content less than 0.5% is termed as low and more than 0.75% as high.

Ortstein : An indurated layer in the B horizon of spodosols in which the cementing material consists of illuviated sesquioxides (mostly iron) and organic matter..

Oxidation ditch : An artificial open channel for partial digestion of liquid organic wastes in which the wastes are circulated and aerated by a mechanical device.

Organic Certification: a marketing label, regulated through the USDA, Agricultural Marketing Service. It specifies that the certified product was grown and processed according to USDA's national organic standards.

Organoarsenicals: (also called **organoarsenic compounds**) compounds that are produced industrially with uses as insecticides, herbicides, and fungicides. In general these applications are declining in step with growing concerns about their impact on the environment and human health.

Organic crop production: refers to an agricultural system that follows the specific, legal requirements outlined in the USDA National Organic Programme (NOP) regulations. For example, no GMOs are permitted and crops are produced without the use of synthetic pesticides or fertilizers. Growers are certified by a USDA-approved certifying agency only after they have demonstrated compliance with stringent NOP standards. http://www.ams.usda.gov/NOP/indexIE.htm

Organic labels Certified organic: is used to label a farm, farmer, or product that has been certified in accordance with USDA National Organic Programme (NOP) regulations. Only farmers that have been inspected and approved by a USDA-accredited organisation (such as the KDA) may sell, label, and represent their products as certified organic. The USDA organic logo may be used on these products.

100% organic: products contain only certified organically produced ingredients (with the exception of salt and water). Producers and handlers must be certified organic to sell, label, or represent their products as 100% organic. The USDA logo may be used on these products.

Organically inclined means the producer prefers organic crop production techniques, but it does not guarantee that organic methods were used exclusively.

Organic:refers to products in which 95% or more of their ingredients are certified organic. Producers and handlers must be certified organic in order to sell, label, or represent their products as organic. The USDA logo may be used on these products.

Made with organic ingredients refers to products that contain certified organic ingredients. At least 70% of the ingredients must be organic; the label may list up to three of these ingredients. Producers and handlers must be certified organic in order to sell, label, or represent their products as "made with organic products." The USDA organic logo may *not* be used on these products.

Permaculture is a contraction of the words "permanent agriculture" and "permanent culture." Based on the work of Australian Bill Bollison, permaculture incorporates techniques from tribal, traditional, and scientific cultures around the world. It is a sustainable form of agriculture that attempts to integrate the production of crops and animals into a low maintenance, balanced system. The three core values are: earth-care, people-care, and fair-share.

Pesticide-free crops are those that have been produced without the use of insecticides, herbicides, fungicides, or rodenticides. Products that bear the Pesticide Free Production logo (trademark of University of Manitoba) have not been treated with pesticides from seedling emergence to market, are non-GMO, and have not been grown where residual pesticides are commercially active.

P

Peat: Unconsolidated soil material consisting largely of undecomposed, or only slightly decomposed, organic matter accumulated under conditions of excessive moisture.

Percolation, soil water : The downward movement of water through soil. Especially, the downward flow of water in saturated or nearly saturated soil at hydraulic gradients of the order of 1.0 or less..

Permeability, soil : The ease with which gases, liquids, or plant roots penetrate or pass through a bulk mass of soil or a layer of soil.

pH : A term used to indicate the degree of acidity or alkalinity, Technically, pH is the common logarithm of the reciprocal of the hydrogen ion concentration of a solution. A pH of 7.0 indicates precise neutrality, higher values indicate increasing alkalinity, and lower values indicate increasing acidity.

Phosphobacterium : A microbial culture containing bacteria which increases the availability of applied and native soil phosphorus. Several species of bacteria are effective for increasing phosphorus availability but *Bacillus magatherium var phosphaticum* is most commonly used. In Russia and several Eastern European countries phosphobacterium culture is sold commercially and is used to inoculate the soil for increasing phosphorus availability.

Phyllosphere : The leaf surface.

Protected Harvest is an independent non-profit organisation that offers a Certified Sustainable label to growers. This program stresses the social and environmental aspects of sustainable agriculture, and includes the use of biointensive IPM. The certification requires a third party audit and an on-site inspection. Standards are specific to the crop and region. http://www.protectedharvest.org

Regenerative agriculture is used to describe a sustainable agricultural system that focuses on restoring soil health and the balance of nature. http://newfarm.rodaleinstitute.org/features/0802/regenerative.shtml

Residue-free: signifies that a product does not have pesticide residues above an established limit set by the producer or company. This label does not mean that pesticides were not used at any time nor does it indicate that the product is 100% free of any chemical residue.

Precision agriculture: It is the application of modern information technologies and makes uses of information system and planning software to provide, process and analyze multi-source data of high spatial and temporal resolution for decision making and operation in the management of crop production.

Productivity, soil: The capacity of a soil for producing a specified plant or sequence of plants under a specified system of management. Productivity emphasizes the capacity of soil to produce crops and should be expressed in terms of yields.

Protein: Any of a group of nitrogen-containing organic compounds formed by the polymerization of a large number of amino acid molecules and that, upon hydrolysis, yield these amino acids. They are essential parts of living matter and are one of the essential food substances of animals.

Puddled soil: Dense, massive soil artificially compacted when wet and having no aggregated structure. The condition commonly results from the tillage of a clayey soil when it is wet.

Putrefiction : A process of degradation by which protein rich material is decomposed under anaerobic conditions thereby releasing foul smelling gases. The final products in putrefaction are ammonia, amines, carbon dioxide, organic acids, hydrogen sulphide etc.

R

Rangeland : Land used for free-grazing livestock.

Recalcitrant : The property of a material to stubbornly resist decomposition. A chemical such as a pesticide that is not decomposed by microorganisms is a recalcitrant pesticide.

Relative humidity : The concentration of water in the air or soil atmosphere relative to the maximum concentration it can hold at the given temperature.

Residual effect of manure : This refers to the residual beneficial effect of application of farmyard manure on the succeeding crops. This beneficial effect is due to improvement in the physical condition of the soil, and also due to the unutilized plant nutrients. It is estimated that only one third of the nitrogen present in farmyard manure is utilized by the first crop. Similarly, about two third of the phosphate is effective but most of the potash is available for the first crop.

Residue conservation : Leaving straw, stubble, trash – crop residues – to rot on or in the ground instead of removing or burning them.

Rhizobia : Bacteria capable of living symbiotically with higher plants, usually in nodules on the roots of legumes, from which they receive their energy, and capable of converting atmospheric nitrogen to combined organic forms; hence, the term *symbiotic nitrogen-fixing bacteria*. (Derived from the generic name *Rhizobium*).

Rhizosphere : The term was introduced by L. Hiltner (1904) to describe that portion of the soil which is in close contact with roots and is influenced by the root system with regard to microbial activity and other related phenomena.

Root exudate : A mixture of organic acids, sugars, and other soluble plant components that escape from roots.

S

Secondary plant nutrients : The secondary plant nutrients are calcium, magnesium and sulphur. These nutrients are called secondary because these are not applied as straight commercial fertilizer but are applied to the soil, indirectly while adding N, P_2O_5 and K_2O in the form of commercial fertilizers. Thus for manufacturers of fertilizers containing major plant nutrients, calcium, magnesium and sulphur are of secondary importance.

Self-mulching soil: A soil in which the surface layer becomes so well aggregated that it does not crust and seal under the impact of rain but instead serves as a surface mulch upon drying.

Septic tank : An underground tank used in the deposition of domestic wastes. Organic matter decomposes in the tank, and the effluent is drained into the surrounding soil.

Sewage sludge: Settled sewage solids combined with varying amounts of water and dissolved materials, removed from sewage by screening, sedimentation, chemical precipitation, or bacterial digestion.

Shelterbelt : A wind barrier of living trees and shrubs established and maintained for protection of farm fields. Syn. Windbreak.

Shifting cultivation : A farming system in which land is cleared, the debris burned, and crops grown for 2-3 years. When the farmer moves on to another plot, the land is then left idle for 5-15 years, then the burning and planting process is repeated.

Sludge: The solid portion of sewage. The sludge is obtained by treating sewage by different methods. Accordingly, sludges of different types are formed. Various types of sludges are settled sludge, digested sludge, activated sludge, digested activatedsludge and chemically precipitated sludge, On an average, sludge contains 1.5 to 3.5 per cent nitrogen, 0.75 to 4.0 per cent P_2O_5 and 0.3 to 0.6 per K_2O.

Soil conditioners : Are chemicals which are added to maintain physical condition of the soil. Such chemicals are polyvinytites polyacrylates, cellulose gums, lignin derivatives and silicates.

Soil conservation: A combination of all management and land-use methods that safeguard the soil against depletion or deterioriation caused by nature and / or humans.

Soil fertility : Soilfertility refers to the inherent capacity of a soil to supply nutrients to plants in adequate amount and in suitable proportion.

Soil organic matter: The organic fraction of the soil that includes plant and animal residues at various stages of decomposition, cells and tissues of soil organisms, and substances synthesized by the soil population. Commonly determined as the amount of organic material contained in a soil sample passed through a 2 mm sieve.

Soil pH : It refers to reaction of the soil as expressed in terms of pH. Soil having pH 7 is neutral and with less than pH 7 is acidic. If the soil pH is more than 7, it is alkaline in reaction.

Soil productivity: Productivity is the present capacity of a soil to produce crop yield under a defined set of management practices. It is measured in terms of the yield in relation to the input of production factor.

Soil water potential (total): A measure of the difference between the free energy state of soil water and that of pure water. Technically it is defined as "that amount of work that must be done per unit quantity of pure water in order to transport reversibly and isothermically an infinitesimal quantity of water from a pool of pure water, at a specified elevation and at atmospheric pressure to the soil water (at the point under consideration)." This total potential consists of the following potentials.

Solid waste: Waste other than sewage, mostly solid. Household garbage is a familiar form of solid waste.

Stomates (or stomata): The controllable openings in the epidermis .

Strip cropping : The practice of growing crops that require different types of tillage, such as row and sod, in alternate strips along contours or across the prevailing direction of wind.

Stubble mulch: The stubble of crops or crop residues left essentially in place on the land as a surface cover before and during the preparation of the seedbed and at least partly during the growing of a succeeding crop.

Sustainable agriculture : According to Consultative Group on International Agricultural Research (CGIAR), it is the successful management of resources to satisfy the changing needs, while maintaining or enhancing the quality of environment and conserving natural resources.or"A sustainable agriculture must be ecologically sound, economically viable, and socially responsible. Furthermore, I contend that these three dimensions of sustainability are inseparable, and thus, are equally critical to long run sustainability." (John Ikerd, U of MO, http://www.sustainable-ag.ncsu.edu/onsustaibableag.htm)

Symbiosis: The living together in intimate association of two dissimilar organisms, the cohabitation being mutually beneficial.

Synthesis: Combination of simple molecules to form another substance e.g., the union of carbon dioxide and water under the action of sunlight in photosynthesis. Adjective from synthesis is synthetic, e.g. synthetic ammonia.

T

Thermophilic organisms : Organisms that grow readily at temperatures above 45°C.

Thiobacillus : Sulphur oxidizing chemoautotrophic bacteria that convert in organic sulphur to sulphate..

U

Urease: A crystalisable protein enzyme that activates hydrolysis of urea.

Uronite: A sugar with a COOH group.

V

Variable charge: An electrical charge on clay or organic matter that changes with changes in soil pH.

Vesicular arbuscular mycorrhiza: A common endomycorrhizal association produced by phycomycetous fungi of the genus *Endogone* and characterised by the development of two types of fungal structures: a) within root cells small structures known as arbuscles and b) between root cells storage organs known as vesicles. Host range includes many agricultural and horticultural crops.

Vesicular pore: A soil pore not connected to other pores.

Virgin soil: A soil that has not been significantly disturbed from its natural environment.

W

Water use efficiency: Dry matter or harvested portion of crop produced per unit of water consumed.

Windbreak: Planting of trees, shrubs, or other vegetation perpendicular, or nearly so to the principal wind direction to protect soils, crops, homesteads, etc., from wind and snow.

X

Xeric : Refersto a dry environment e.g. desert.

Xerophytes : Plants that grow in or on extremely dry soils or soil materials.

Y

Yield : The individuals or biomass removed when the plant population is harvested.

Yield potential: The total production capacity of a crop.